PROGRESS TOWARD RESTORING THE EVERGLADES
The Seventh Biennial Review - 2018

Committee on Independent Scientific Review of
Everglades Restoration Progress

Water Science and Technology Board

Board on Environmental Studies and Toxicology

Division on Earth and Life Studies

A Consensus Study Report of

The National Academies of
SCIENCES • ENGINEERING • MEDICINE

THE NATIONAL ACADEMIES PRESS
Washington, DC
www.nap.edu

THE NATIONAL ACADEMIES PRESS   500 Fifth Street, NW   Washington, D.C. 20001

Support for this study was provided by the Department of the Army under Cooperative Agreement No. W912EP-04-2-0001. Support for this project was also provided by the U.S. Department of the Interior and the South Florida Water Management District. Any opinions, findings, conclusions, or recommendations expressed in this publication do not necessarily reflect the views of any organization or agency that provided support for the project.

International Standard Book Number-13: 978-0-309-47978-3
International Standard Book Number-10: 0-309-47978-9
Digital Object Identifier: https://doi.org/10.17226/25198

Cover credit: David Policansky

Additional copies of this report are available for sale from the National Academies Press, 500 Fifth Street, NW, Keck 360, Washington, DC 20001; (800) 624-6242 or (202) 334-3313; http://www.nap.edu/.

Copyright 2018 by the National Academy of Sciences. All rights reserved.

Printed in the United States of America

Suggested citation: National Academies of Sciences, Engineering, and Medicine. 2018. *Progress Toward Restoring the Everglades: The Seventh Biennial Review—2018*. Washington, DC: The National Academies Press. doi: https://doi.org/10.17226/25198.

The National Academies of
SCIENCES · ENGINEERING · MEDICINE

The **National Academy of Sciences** was established in 1863 by an Act of Congress, signed by President Lincoln, as a private, nongovernmental institution to advise the nation on issues related to science and technology. Members are elected by their peers for outstanding contributions to research. Dr. Marcia McNutt is president.

The **National Academy of Engineering** was established in 1964 under the charter of the National Academy of Sciences to bring the practices of engineering to advising the nation. Members are elected by their peers for extraordinary contributions to engineering. Dr. C. D. Mote, Jr., is president.

The **National Academy of Medicine** (formerly the Institute of Medicine) was established in 1970 under the charter of the National Academy of Sciences to advise the nation on medical and health issues. Members are elected by their peers for distinguished contributions to medicine and health. Dr. Victor J. Dzau is president.

The three Academies work together as the **National Academies of Sciences, Engineering, and Medicine** to provide independent, objective analysis and advice to the nation and conduct other activities to solve complex problems and inform public policy decisions. The National Academies also encourage education and research, recognize outstanding contributions to knowledge, and increase public understanding in matters of science, engineering, and medicine.

Learn more about the National Academies of Sciences, Engineering, and Medicine at **www.nationalacademies.org**.

The National Academies of
SCIENCES • ENGINEERING • MEDICINE

Consensus Study Reports published by the National Academies of Sciences, Engineering, and Medicine document the evidence-based consensus on the study's statement of task by an authoring committee of experts. Reports typically include findings, conclusions, and recommendations based on information gathered by the committee and the committee's deliberations. Each report has been subjected to a rigorous and independent peer-review process and it represents the position of the National Academies on the statement of task.

Proceedings published by the National Academies of Sciences, Engineering, and Medicine chronicle the presentations and discussions at a workshop, symposium, or other event convened by the National Academies. The statements and opinions contained in proceedings are those of the participants and are not endorsed by other participants, the planning committee, or the National Academies.

For information about other products and activities of the National Academies, please visit www.nationalacademies.org/about/whatwedo.

COMMITTEE ON INDEPENDENT SCIENTIFIC REVIEW OF EVERGLADES RESTORATION PROGRESS

WILLIAM G. BOGGESS, *Chair*, Oregon State University, Corvallis
MARY JANE ANGELO, University of Florida, Gainesville
CHARLES T. DRISCOLL, Syracuse University, New York
M. SIOBHAN FENNESSY, Kenyon College, Gambier, Ohio
WENDY D. GRAHAM, University of Florida, Gainesville
KARL E. HAVENS, University of Florida, Gainesville
FERNANDO R. MIRALLES-WILHELM, University of Maryland, College Park
DAVID H. MOREAU, University of North Carolina, Chapel Hill
GORDON H. ORIANS, University of Washington, Seattle
DENISE J. REED, University of New Orleans, Louisiana
JAMES E. SAIERS, Yale University, Connecticut
ERIC P. SMITH, Virginia Polytechnic Institute and State University, Blacksburg
DENICE H. WARDROP, Pennsylvania State University, University Park
GREG D. WOODSIDE, Orange County Water District, Fountain Valley, California

Staff

STEPHANIE E. JOHNSON, Study Director, Water Science and Technology Board
BRENDAN R. McGOVERN, Research Assistant, Water Science and Technology Board
DAVID J. POLICANSKY, Scholar, Board on Environmental Studies and Toxicology

Acknowledgments

Many individuals assisted the committee and the National Academies of Sciences, Engineering, and Medicine staff in their task to create this report. We would like to express our appreciation to the following people who provided presentations or public comment to the committee or served as field trip guides.

Nick Aumen, U.S. Geological Survey
James Beerens, U.S. Geological Survey
Laura Brandt, U.S. Fish and Wildlife Service
Tim Breen, U.S. Fish and Wildlife Service
Trisston Brown, U.S. Army Corps of Engineers
Rich Budell, Budell Water Group
Cara Capp, National Parks Conservation Association
Dean Carpenter, Albemarle-Pamlico National Estuary Partnership
Bill Causey, National Oceanic and Atmospheric Administration
Bahram Charkhian, South Florida Water Management District
Cris Costello, Sierra Club
Dan Crawford, U.S. Army Corps of Engineers
Steve Culberson, Delta Stewardship Council
Steve Davis, Everglades Foundation
Celeste De Palma, Audubon Florida
Michael Drog, U.S. Army Corps of Engineers
Dennis Duke, U.S. Geological Survey
Gretchen Ehlinger, U.S. Army Corps of Engineers
Shannon Estenoz, Department of Interior
Michelle Ferree, South Florida Water Management District
Brad Foster, U.S. Army Corps of Engineers
Jim Fourqurean, Florida International University
Tom Frankovich, Florida International University
Evelyn Gaiser, Florida International University

Donna George, U.S. Army Corps of Engineers
Alex Gillen, Bull Sugar
David Gillings, Palm Beach County
Howie Gonzales, U.S. Army Corps of Engineers
Patti Gorman, South Florida Water Management District
Susan Gray, South Florida Water Management District
Paul Gray, Audubon Florida
Tim Gysan, U.S. Army Corps of Engineers
Chuck Hanlon, South Florida Water Management District
Rainer Hoenicke, Delta Stewardship Council
Bud Howard, Loxahatchee River District
Tom James, South Florida Water Management District
LTC Jennifer Reynolds, U.S. Army Corps of Engineers
Kang-Ren Jin, South Florida Water Management District
Bob Johnson, U.S. National Park Service
Paul Julian, Florida Department of Environmental Protection
Kelly Keefe, U.S. Army Corps of Engineers
Chris Kelble, National Oceanic and Atmospheric Administration
William "Chad" Kennedy, Florida Department of Environmental Protection
Kevin Kotun, U.S. National Park Service
Glenn Landers, U.S. Army Corps of Engineers
Jennifer Leeds, South Florida Water Management District
Andy LoSchiavo, U.S. Army Corps of Engineers
Ernie Marks, South Florida Water Management District
Jenna May, U.S. Army Corps of Engineers
Agnes McLean, U.S. National Park Service
Miles Meyer, U.S. Fish and Wildlife Service
Brenda Mills, South Florida Water Management District
June Mirecki, U.S. Army Corps of Engineers
Robert Mooney
Matt Morrison, South Florida Water Management District
Melissa Nasuti, U.S. Army Corps of Engineers
Mark Nelson, Jonathan Dickinson State Park
Mindy Parrott, South Florida Water Management District
April Patterson, U.S. Army Corps of Engineers
Mark Perry, Everglades Coalition
Patrick Pitts, U.S. Fish and Wildlife Service
Rene Price, Florida International University
Bob Progulske, U.S. Fish and Wildlife Service
Jed Redwine, U.S. National Park Service

Gregg Reynolds, U.S. National Park Service
Stephanie Romanach, U.S. Geological Survey
Barry Rosen, U.S. Geological Survey
Rob Rossmanith, Jonathan Dickinson State Park
David Rudnick, U.S. National Park Service
Steve Schubert, U.S. Fish and Wildlife Service
Dawn Shirreffs, Everglades Foundation
Fred Sklar, South Florida Water Management District
Janet Starnes, South Florida Water Management District
Eric Summa, U.S. Army Corps of Engineers
Donatto Surratt, U.S. National Park Service
Peter Tango, U.S. Geological Survey
Kim Taplin, U.S. Army Corps of Engineers
Brett Thomas, U.S. Army Corps of Engineers
Joel Trexler, Florida International University
Tiffany Troxler, Florida International University
Diana Umpierre, Sierra Club
Stuart Van Horn, South Florida Water Management District
Craig van der Heiden, Miccosukee Tribe
Eva Velez, South Florida Water Management District
Bob Verrastro, South Florida Water Management District
Zach Welch, South Florida Water Management District
Walter Wilcox, South Florida Water Management District
Mike Yustin, Martin County

Preface

South Florida is blessed with a unique, wonderfully diverse, and geographically extensive wetland ecosystem reaching from just south of Orlando to the Florida Keys. After nearly 150 years of drainage, channelization, and flood control actions, this extraordinary natural resource has been dramatically altered and continues to decline. Where water once traveled slowly south toward the Everglades National Park through ridge and slough wetlands, marl prairies, and sawgrass plains, it is now often diverted to the ocean or to other uses—less than half reaches its historic destination. The quality of the water remaining in the system is compromised by the phosphorus, nitrogen, mercury, and other contaminants introduced by urban development, agriculture, and industry. The combination of reduced water flow and degraded water quality impacts has adversely changed land formation and vegetation patterns. Experts recognized more than 20 years ago that significant action was needed to rescue and preserve this national treasure.

The U.S. Congress authorized the Comprehensive Everglades Restoration Plan (CERP) in 2000 as the multidecadal, multibillion-dollar response. The CERP is focused on restoring, preserving, and protecting the South Florida ecosystem while providing for other water-related needs of the region. This massive restoration program, the largest in U.S. history, is jointly administered by the U.S. Army Corp of Engineers (USACE) and the South Florida Water Management District (SFWMD) and is equally funded by federal and Florida monies. As part of the initial authorization, Congress mandated periodic independent reviews of progress toward restoration of the Everglades natural system. The National Academies of Sciences, Engineering, and Medicine's Committee on Independent Scientific Review of Everglades Restoration Progress, or CISRERP, was formed for this purpose in 2004. This report represents the seventh biennial review of CERP progress by this committee.

This seventh iteration of CISRERP includes a mix of science and engineering specialists brought together for their combined expertise in environmental, biological, hydrologic, and geographic sciences; systems engineering; project and program administration; law; economics; and public policy. These experts were selected for their eminence in their fields, as well as their experience with complex, natural systems similar to the Everglades. The committee met five times over a 14-month period, including four times in Florida. We reviewed a large volume of written material and heard oral presentations from state and federal agency personnel, academic researchers, interest groups, and members of the public. The committee's task is a daunting one, given the size and complexity of the Everglades ecosystem and corresponding scope of the CERP. I greatly appreciate the time, attention, and thought each committee member invested in understanding this complex system. I also appreciate the careful, rigorous analyses, expert judgment, constructive comments and reviews, and good humor with which they conducted their work. The report presents our consensus view of restoration accomplishments and challenges that have emerged during not only the past 2 years but also the nearly two decades since the project was authorized.

The committee thanks many individuals for the information and resources they provided. Specifically, we appreciate the efforts of the committee's technical liaisons—David Tipple (USACE), Donna George (USACE), Glenn Landers (USACE), Rod Braun (SFWMD), Megan Jacoby (SFWMD), and Robert Johnson (Department of the Interior)—who responded to numerous information requests and facilitated the committee's access to agency resources and expertise when needed. The committee is also grateful to the numerous individuals who shared their insights and knowledge of Everglades restoration through presentations, field trips, and public comments (see Acknowledgments).

The committee had the good fortune to be assisted by three dedicated and very talented National Academies' staff: Stephanie Johnson, David Policansky, and Brendan McGovern. Serving as senior project officer for all seven CISRERP panels, Stephanie Johnson orchestrated the study for the National Academies. Her comprehensive understanding of CERP and its component parts, the complex physical system, agency interrelationships, diverse constituencies, and the surrounding political landscape gave her an unparalleled vantage point in supporting the committee's activities. Stephanie's stewardship of the final report creation process, initial drafting through completion, was exceptional. National Academies of Sciences, Engineering, and Medicine scholar David Policansky is also a veteran of all seven CISRERP panels, and his experience, insightful observations, and illuminating questions were fundamental to the committee's deliberations. Brendan McGovern most ably supported the logistical needs of the committee. Brendan was also a valued contributor in completing the final report. Simply put, this report would not have been possible without the National

Academies staff's exceptional support and good humor. I know I speak for the entire committee in expressing our profound respect and appreciation.

This Consensus Study Report was reviewed in draft form by individuals chosen for their diverse perspectives and technical expertise. The purpose of this independent review is to provide candid and critical comments that will assist the National Academies of Sciences, Engineering, and Medicine in making each published report as sound as possible and to ensure that it meets the institutional standards for quality, objectivity, evidence, and responsiveness to the study charge. The review comments and draft manuscript remain confidential to protect the integrity of the deliberative process.

We thank the following individuals for their review of this report:

Mary Christman, MCC Statistical Consulting LLC, Gainesville, FL
Peter Goodwin, University of Maryland Center for Environmental Science, Cambridge
Matthew Harwell, U.S. Environmental Protection Agency, Gulf Breeze, FL
Carl Hershner, Virginia Institute of Marine Science, Gloucester Point
Rainer Hoenicke, Delta Stewardship Council, Sacramento, CA
John Kominoski, Florida International University, Miami
Dorothy Merritts, Franklin & Marshall College, Lancaster, PA
Jayantha Obeysekera, Florida International University, Miami
William Schlesinger (NAS), Cary Institute of Ecosystem Studies (*retired*), Millbrook, NY
Alan Steinman, Grand Valley State University, Allendale, MI
Kirsten Work, Stetson University, DeLand, FL

Although the reviewers listed above provided many constructive comments and suggestions, they were not asked to endorse the conclusions or recommendations of this report nor did they see the final draft before its release. The review of this report was overseen by Bonnie McCay, Rutgers University, and Kenneth Potter, University of Wisconsin-Madison. They were responsible for making certain that an independent examination of this report was carried out in accordance with the standards of the National Academies and that all review comments were carefully considered. Responsibility for the final content rests entirely with the authoring committee and the National Academies.

In this seventh CISRERP review cycle, our committee has the pleasure of reporting on the early ecosystem benefits from CERP investments. The past 2 years have also been marked by impressive progress in meeting water quality targets, construction, and project planning. Another portion of our charge is to evaluate the effectiveness of the monitoring and assessment program in supporting restoration efforts. In this report, we provide a detailed review of CERP project-level monitoring and assessment with an eye toward improving the efficiency and effectiveness of the CERP monitoring program within existing resource constraints.

A third part of our charge is to illuminate issues that may impede or diminish the overall success of CERP. In the past, we have highlighted the slow rate of program implementation, the focus on the periphery rather than the center, adverse trajectories for natural system components, potential impacts of climate change, implications of invasive species, and the need for a CERP update. We believe our independent reviews have brought an important and timely focus on these critical concerns. In this review we turn our attention to the future. During the past 30 years of Everglades restoration, the past has been prologue. Understanding the past tells us what made this ecosystem unique and special, including the processes that created and sustained it, informing the restoration efforts. The original CERP plan was formulated based on a pre-drainage or early-twentieth century vision of the historical Everglades and past sea levels and rainfall and temperature distributions. But the past is not prologue for the future environment of South Florida. There is now ample evidence that rainfall and temperature distributions in South Florida are changing and compelling recent evidence that sea-level rise in South Florida is accelerating. It is clear that the Greater Everglades of 2050 and beyond will be much different from what was envisioned at the time of the CERP conceptual plan, known as the Yellow Book. These changes have profound implications for the interrelated challenges of restoring the natural system, providing flood protection, and meeting the water demands of a growing population. Everglades restoration has always been an ambitious and complex endeavor; our current review emphasizes how it is also dynamic and the importance of focusing restoration on the future Everglades, rather than on the past Everglades. We offer this report with an eye to that future and in support of that grand endeavor.

William Boggess, *Chair*
Committee on Independent Scientific Review of
Everglades Restoration Progress (CISRERP)

Contents

ACRONYMS		xvii
SUMMARY		1
1	INTRODUCTION	13
2	THE RESTORATION PLAN IN CONTEXT	21
3	RESTORATION PROGRESS	35
4	MONITORING AND ASSESSMENT	101
5	LAKE OKEECHOBEE REGULATION	133
6	A CERP MID-COURSE ASSESSMENT: SUPPORTING SOUND DECISION MAKING FOR THE FUTURE EVERGLADES	159
REFERENCES		189

APPENDIXES
A	The National Academies of Sciences, Engineering, and Medicine Everglades Reports	205
B	Water Science and Technology Board and the Board on Environmental Studies and Toxicology	213
C	Biographical Sketches of Committee Members and Staff	215

Acronyms

AF	acre-feet
ASR	aquifer storage and recovery
BACI	before-after control-impact
BBCW	Biscayne Bay Coastal Wetlands
BMP	best management practice
CEPP	Central Everglades Planning Project
CERP	Comprehensive Everglades Restoration Plan
CESI	Critical Ecosystem Studies Initiative
CISRERP	Committee on Independent Scientific Review of Everglades Restoration Progress
CROGEE	Committee on Restoration of the Greater Everglades Ecosystem
C&SF	Central and Southern Florida
DMDU	decision making under deep uncertainty
DOI	U.S. Department of the Interior
DPM	Decomp(artmentalization) Physical Model
EAA	Everglades Agricultural Area
EDRR	early detection and rapid response
EPA	U.S. Environmental Protection Agency
ERTP	Everglades Restoration Transition Plan
FEB	flow equalization basin
FISK	Fish Invasiveness Screening Kit
FY	fiscal year
GCM	general circulation model

HHD	Herbert Hoover Dike	
IDS	Integrated Delivery Schedule	
IRL-S	Indian River Lagoon-South	
LNWR	Loxahatchee National Wildlife Refuge	
LOEM	Lake Okeechobee Environment Model	
LORS	Lake Okeechobee Regulation Schedule	
LTER	Long-term Ecological Research	
MAP	monitoring and assessment plan	
NASEM	National Academies of Sciences, Engineering, and Medicine	
NDVI	Normalized Difference Vegetation Index	
NGVD	National Geodetic Vertical Datum	
NOAA	National Oceanic and Atmospheric Administration	
NPS	National Park Service	
NRC	National Research Council	
PPA	project partnership agreement	
ppb	parts per billion	
RCP	representative concentration pathway	
RDM	robust decision making	
RECOVER	REstoration, COordination, and VERification	
RPA	reasonable and prudent alternative	
RSM	Regional Simulation Model	
SAV	submerged aquatic vegetation	
SFERTF	South Florida Ecosystem Restoration Task Force	
SFWMD	South Florida Water Management District	
SFWMM	South Florida Water Management Model	
SSR	System Status Report	
STA	stormwater treatment area	
TMDL	total maximum daily load	
USACE	U.S. Army Corps of Engineers	
USDA	U.S. Department of Agriculture	

WAI	wetland affinity index
WCA	Water Conservation Area
WERP	Western Everglades Restoration Project
WQBEL	water quality–based effluent limit
WRDA	Water Resources Development Act
WSE	Water Supply and Environment
WY	water year (May 1 to April 30)

Summary

During the past century, the Everglades, one of the world's treasured ecosystems, has been dramatically altered by drainage and water management infrastructure that was intended to improve flood management, urban water supply, and agricultural production. The remnants of the original Everglades now compete for water with urban and agricultural interests and are impaired by contaminated runoff from these two sectors. The Comprehensive Everglades Restoration Plan (CERP), a joint effort launched by the state and the federal government in 2000, seeks to reverse the decline of the ecosystem. The multibillion-dollar project was originally envisioned as a 30- to 40-year effort to achieve ecological restoration by reestablishing the natural hydrologic characteristics of the Everglades, where feasible, and to create a water system that serves the needs of both the natural and the human systems of South Florida.

The National Academies of Sciences, Engineering, and Medicine established the Committee on Independent Scientific Review of Everglades Restoration Progress in 2004 in response to a request from the U.S. Army Corps of Engineers (USACE), with support from the South Florida Water Management District (SFWMD) and the U.S. Department of the Interior (DOI), based on Congress's mandate in the Water Resources Development Act of 2000 (WRDA 2000). The committee is charged to submit biennial reports that review the CERP's progress in restoring the natural ecosystem. This is the committee's seventh report. Each report provides an update on natural system restoration progress during the previous 2 years, describes substantive accomplishments (Chapter 3), and reviews developments in research, monitoring, and assessment that inform restoration decision making (Chapter 4 and 6). In each new report, the committee also identifies issues for in-depth evaluation considering new CERP program developments, policy initiatives, or improvements in scientific knowledge that have implications for restoration progress (see Chapter 1 for the committee's full statement of task). For the 2018 review, the committee performed an in-depth review of CERP monitoring, with particular emphasis on project-level monitoring and assessment (Chapter 4). To inform forward-looking systemwide planning decisions, the committee synthesized recent information on Lake Okeechobee and the effects of water levels on lake ecology (Chapter 5) and reexamined the value of a mid-course assessment of the CERP outcomes focused on the South Florida ecosystem of the future (Chapter 6).

OVERALL EVALUATION OF PROGRESS AND CHALLENGES

During the past 2 years, there have been impressive efforts toward project planning associated with four new projects. A vision for planned CERP storage, at least in the northern portion of the system, is now becoming clear, although the future storage to be provided by Lake Okeechobee remains unresolved. Recent analysis has shown that coordination of operations can make more effective use of available water, potentially reducing the amount of CERP storage needed to achieve successful restoration. However, the systemwide implications of the new projects, which have been in planning concurrently, have not been assessed. Construction continues on five CERP projects (Figure S-1), and state funding for CERP project construction has increased, while two major non-CERP projects have been completed. Documentation and analysis of incremental restoration benefits from project implementation to date have been inadequate, primarily because of limitations in project-level monitoring and assessment efforts. Improvements to the monitoring and assessment program, at both project and systemwide scales, are recommended to increase the usefulness of monitoring data for CERP decision makers.

Eighteen years into the CERP, the committee recommends a mid-course assessment that analyzes projected CERP outcomes in the context of future stressors. Rather than continuing its primary focus on restoring pre-drainage conditions and basing decisions on the ability to achieve those conditions under contemporary climate (1965-2005), the CERP program should emphasize restoration focused on the future of the South Florida ecosystem and build upon the accumulating knowledge base to support successful implementation of this program. This effort requires a rigorous assessment of the latest CERP project plans that examines their integrated performance under future climate and sea level–rise scenarios and other stressors. With seven large projects authorized and awaiting appropriations for construction and three additional projects nearing the end of their planning processes, the time is right for a mid-course assessment. This information could then inform robust decisions about future planning, funding, sequencing, and adaptive management. Implementing a restoration program that is resilient to future conditions also requires a science program that can bring

Summary 3

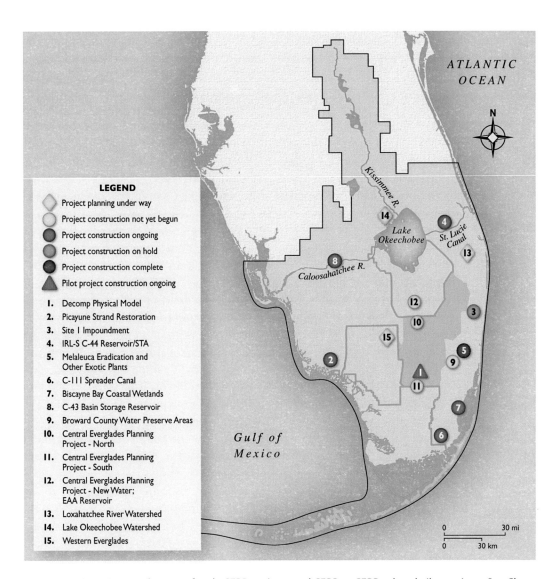

FIGURE S-1 Locations and status of early CERP projects and CERP or CERP-related pilot projects. See Chapter 3 for more information on CERP implementation progress.

SOURCE: © International Mapping Associates.

the latest information and tools into CERP planning and implementation. The major conclusions and recommendations of the report are summarized below.

RESTORATION PROGRESS

In Chapter 3, the committee outlines the major accomplishments of restoration, with an emphasis on natural system restoration progress, and discusses issues that may impact progress. CERP project implementation remains in the early stages. If recent (5-year average) federal funding levels continued and were matched by the state, construction of the remaining components of the congressionally authorized projects could take approximately 65 years; construction of projects in planning or those currently unplanned would further lengthen that timeline. At this pace of restoration, it is even more imperative that agencies anticipate and design for the Everglades of the future.

Incremental restoration progress from early CERP projects is difficult to evaluate because of a lack of rigorous assessment of outcomes relative to project goals and some limitations in existing monitoring plans. The committee reviewed available data and analysis on the restoration progress associated with three early CERP projects in which substantial project components are now in place and operating (see Figure S-1, Nos. 2, 6, and 7). The Picayune Strand Restoration Project shows increased water levels in the area of the two canals plugged to date. Hydrologic conditions are expected to improve further toward conditions at the reference sites once neighboring canals are plugged. Some early indicators of habitat response at Picayune Strand are apparent in the species composition of groundcover vegetation and suppression of some exotic species, but other ecological indicators, such as increased cypress regeneration, have not shown significant change. This lack of response could be due to lag times in ecological response, limitations in the monitoring plan, or insufficient hydrologic restoration to date. Analysis of these or other factors is an essential but missing component of performance assessment. At the C-111 Spreader Canal project, neither hydrologic nor ecological response in Taylor Slough or Florida Bay due to the project has been documented based on monitoring data because the monitoring and assessment plans are not robust enough to discern project impacts from existing hydrologic variability. The lack of specific numeric targets and an explicit plan and model to evaluate restoration progress hinders restoration assessment of these two projects. The Biscayne Bay Coastal Wetlands monitoring program has documented hydrologic and ecological responses, although both are limited by the small spatial scale of the components that have been implemented and important project components that are not yet constructed.

Concurrent project planning efforts have significantly advanced the CERP

vision for water storage, but a holistic understanding of the benefits of the combined projects at a systemwide scale and their resilience to sea-level rise and climate change is lacking. Tentatively selected plans have been developed for the Everglades Agricultural Area (EAA) Reservoir and the Lake Okeechobee Watershed Restoration Project, which together propose adding 283,000 acre-feet (AF) of surface storage and 80 aquifer storage and recovery wells. Each project is expected to reduce high-volume discharges from Lake Okeechobee to the Northern Estuaries and to modestly improve the period that Lake Okeechobee stage is at ecologically preferred levels. The EAA Reservoir also provides moderate hydrologic improvements to Water Conservation Area (WCA) 2A and northern WCA-3A. By 2019, all of the large CERP storage projects at the northern end of the system will have been planned, with only Lake Okeechobee and southern storage (i.e., Lake Belt) remaining unresolved. Preliminary modeling suggests that with system optimization, the full storage planned in the original CERP may not be needed to provide the flows into the northern end of the Everglades as envisioned in the CERP. However, a new integrated, systemwide modeling of the planned projects is needed to understand the combined benefits relative to restoration objectives. More rigorous analysis of the potential effects of climate change and sea-level rise on restoration outcomes is necessary in planning for all projects, so that restoration investments are designed for and more resilient to future conditions. The SFWMD and the Interagency Modeling Center have the talent and tools to conduct these analyses, and the SFWMD is pursuing this approach for planning and management issues outside of the CERP.

Impressive advances have been made toward water quality objectives in the stormwater treatment areas (STAs). The lowest flow-weighted mean total phosphorus concentrations to date (15 ppb for all STAs combined) were attained in water year 2017, and continued water quality treatment and science investments through the Restoration Strategies program are expected to further reduce phosphorus levels toward the 13 ppb goal. Achieving this goal is a necessary step to move forward with new water flows in the central Everglades. Understanding the dynamic ecological responses to restored flows (and the relative importance of phosphorus concentration and load in controlling ecosystem response) during these transitions is an emerging challenge. Where existing flows are currently being redistributed, as in the Decomp Physical Model and the non-CERP Florida Bay Initiative, project teams are following adaptive management approaches where feasible to learn from these efforts and to inform future Everglades flow restoration projects.

The recent completion of two major non-CERP projects is expected to provide important restoration benefits to Everglades National Park and increasing operational flexibility for managing high water events throughout the remnant

Everglades. Completion of the Modified Water Deliveries to Everglades National Park (Mod Waters) and C-111 South Dade projects in August 2018 are major achievements that have been more than 25 years in the making. Development of the Combined Operational Plan is under way, which will quantify the benefits provided by these projects.

MONITORING AND ASSESSMENT

Monitoring is essential to assess the effectiveness of ecosystem restoration efforts (i.e., what was the response?) and support adaptive management (i.e., if the expected outcomes did not occur, why not?). The collection and assessment of monitoring data are necessary to communicate the outcomes of restoration efforts to decision makers and the public, support learning from the restoration outcomes, and guide decisions about future changes that may be needed. The committee's conclusions and recommendations for monitoring were informed by a review of project-level monitoring for three early CERP projects (Picayune Strand, Biscayne Bay Coastal Wetlands [Phase 1], and C-111 Spreader Canal Western) and of the CERP systemwide monitoring program. Although this and previous National Academies committees have recommended improvements in CERP-associated monitoring programs, this does not necessarily mean that additional funding for monitoring is required. There are many ways to improve both the efficiency and the effectiveness of the CERP monitoring program within the existing monitoring budget.

The three CERP projects analyzed vary in the extent to which they have implemented effective monitoring plans. The RECOVER 2006 Assessment Strategy for the Monitoring and Assessment Plan provides valuable guidance on how to establish monitoring plans to detect change and evaluate progress toward goals. However, the three projects reviewed in Chapter 4 have not implemented this guidance systematically. For example, there is variation in the extent to which quantitative restoration objectives are articulated. Not all projects have established a clear sampling design and data analysis plan as part of the monitoring plan, which could limit the usefulness of the results.

Quantitative restoration objectives, with accompanying expectations of how and when they will be achieved by management actions, should be developed for each project during the project development process. Quantitative objectives are needed to effectively measure restoration progress and operationalize goals. In addition, an acceptable level of variability of monitoring data around these objectives should be established so that management actions can be adjusted and adapted if the desired outcome is not being achieved. In the early stages of project development, project teams may be more comfortable with narrative objectives. However, it is essential to establish quantitative objectives as part of the monitoring plan with uncertainty described as appropriate. As programs evolve, more is learned about project functioning, and knowledge and modeling tools improve, the quantitative objectives can be refined.

Monitoring plans should include an evaluation of the ability to detect restoration success given natural variability and sampling constraints. Models and historic monitoring data can be used to select metrics and design sampling plans to determine restoration success with a high degree of certainty, considering natural variability, expected changes from factors such as sea-level rise, and constraints such as site accessibility, funding, and personnel. These analyses should help direct monitoring investments to where they will be most effective.

Modeling and statistical tools should be used in combination with monitoring data to assess restoration performance. External factors, such as precipitation and temperature variability, impact hydrologic and ecological responses, making it difficult to determine ecosystem response to restoration projects when compared to baseline data. Where feasible, reference and control sites can be used to quantify project-related effects, but for most Everglades projects, well-characterized reference and control sites are not available. Additional tools, such as modeling and statistical analyses, are essential to help quantify the effects of the projects and to separate them from ongoing system variability and trends. Modeling tools can be used to separate the effects of other long-term changes, such as sea-level rise or invasive species, on project performance as well as to understand the effects of an individual project within a region that is affected by multiple, interacting projects. Although the CERP has a strong modeling program for project planning, models are rarely used to interpret monitoring data, greatly reducing the potential value of existing data. When numerical or statistical models are to be used in performance assessment, the data analysis plan should be identified before the data are collected to reduce bias in the assessment.

Project-level monitoring should be revisited periodically to ensure that sampling designs and data-analysis plans are effective and efficient and that monitoring investments yield useful information. Periodic reviews would include assessing the usefulness of the monitoring data to meet decision-making needs and the relevance of the selected indicators to the questions being asked. Other considerations include the validity of the conceptual model, the timing and rate of ecosystem response relative to sampling intervals, the adequacy of the spatial scale of monitoring considering the scale of anticipated response, and the use of rigorous computational or statistical tools for data analysis. Such reevaluation should lead to more effective and efficient performance monitoring and will strengthen the capacity to learn through adaptive management.

The full implementation of adaptive management plans will substantially

increase learning about the restoration process. Adaptive management allows learning to take place as new knowledge is gained about ecosystem response to restoration and how changing future conditions (e.g., climate change, sea-level rise) might affect restoration outcomes. Only one of the three CERP projects analyzed (Biscayne Bay Coastal Wetlands) has an established adaptive management plan. Without an adaptive management plan, it is difficult to structure monitoring and evaluation so that new knowledge can be applied in a flexible decision-making process. Performance monitoring may show that project objectives are not being met, but performance monitoring alone cannot explain the reasons for failure or inform restoration decisions. Learning through monitoring is also limited by the lack of integration of modeling with monitoring, which can aid in setting quantitative objectives and projecting reference conditions. Monitoring plans for adaptive management should evaluate whether the restoration project is expected to result in measurable change with high certainty for adaptive management indicators and over what time frame.

The CERP program currently lacks a mechanism for multiagency assessment and reporting of project-level monitoring results. The RECOVER System Status Reports (SSRs) provide comprehensive multiagency analysis and synthesis of systemwide monitoring and assessment of trends, but they do not provide analysis and assessments of individual project performance. Currently, most reporting of project-level monitoring data occurs via the South Florida Environmental Reports (SFERs), which annually compile the data associated with permit monitoring. However, these reports contain limited analysis of long-term trends, project performance relative to expected objectives, and potential adaptive management needs. Additionally, the SFERs do not provide the opportunity for multiagency perspectives or RECOVER input. A variety of other reports, many by contractors, also provide sometimes fragmented summaries of data from monitoring but information on overall project performance relative to objectives remains lacking. A better-organized, multiagency analysis and assessment of project performance based on monitoring results should be developed to provide transparency to decision makers, funders, and stakeholders. This effort would also help support project-level adaptive management efforts.

The upcoming RECOVER review of its systemwide monitoring plan should be embraced as an opportunity to improve its effectiveness and efficiency. Many of the same issues addressed in project-level monitoring, such as the ability of the sampling plan to address the key questions and the availability of data to allow adaptation of management actions if the desired outcomes are not being achieved, are evident in current approaches to systemwide monitoring. The monitoring review, scheduled to begin in 2019, should also consider the relevance and usefulness of indicators, statistical rigor of the assessment, use of modeling for data analysis, and the appropriateness of the spatial and temporal sampling design to ensure that the investments in monitoring are being made toward data that can inform assessments and decision making. Scientists should understand and incorporate the needs of decision makers into the monitoring program. Similarly, decision makers should understand what information scientists can and cannot provide. This approach will require an iterative two-way dialogue between managers and scientists covering such issues as risk tolerance or aversion, what amount of confidence in data summaries is acceptable and possible, which indicators are most important and feasible to monitor, what decisions the information will be used for, and what information is of most scientific value for specific decisions. The process by which monitoring reviews are performed requires a thoughtful and intentional approach, such as the inclusion of stakeholders, modelers, and independent monitoring experts in the review process. Periodic systemwide reviews of monitoring should be incorporated into the work plan of RECOVER so that the monitoring program remains effective and appropriate in the years ahead.

LAKE OKEECHOBEE REGULATION

Lake Okeechobee is the last major component of water storage in the northern end of the South Florida ecosystem to be resolved, and its regulation schedule has significant implications for conditions throughout the ecosystem. The lake regulation schedule will soon be revisited to determine new operational rules. The completion of the Herbert Hoover Dike rehabilitation project could enable higher water levels to be held within Lake Okeechobee, although the feasibility of higher water levels must still be determined through an updated risk assessment. The regulation schedule revision process also considers tradeoffs among the ecological conditions in the lake, the Northern Estuaries, and the Everglades, as well as water supply and flood management. Hydrologic and ecological modeling tools have been developed to assess potential benefits and impacts from various regulation schedules on the lake and broader region. To inform that process and in response to frequent questions about the impacts of increased water levels on the ecology of Lake Okeechobee, the committee summarized the latest information and identified key research needs to help inform the within-lake portion of the tradeoff analysis.

The magnitude of ecological impacts in the lake from additional storage will depend upon antecedent ecological conditions. Improved understanding of these dependencies could be used to inform real-time operations to reduce adverse ecological effects and provide more flexibility given appropriate risk tolerance in lake management. A new regulatory schedule that stores more water in Lake Okeechobee would require tradeoffs between in-lake ecological impacts and ecological and water supply benefits throughout the South Florida ecosystem. Past research has shown that ecological conditions in the lake are adversely affected by high water levels (above ~16 feet) and multiple consecutive years without low water levels (~12 feet). Additionally, reversals of water level recession during spring nesting can adversely affect wading birds and snail kites. However, there are considerable uncertainties about high water impacts to submerged aquatic vegetation (SAV) and near-shore emergent vegetation, which provide important ecological services in the lake, because many of the effects of high water depend on antecedent conditions. For example, high stage effects on SAV vary depending upon whether the plants are healthy and mature, stressed, or just recovering after a prior impact. Reducing those uncertainties and using that information to inform operations could reduce the ecological impacts associated with increased storage.

Adjustments to Lake Okeechobee monitoring and full integration of modeling tools would provide rigorous science-based information to support a regulation schedule review and real-time optimization of operations under any regulation schedule. Refinements to the ecological monitoring and adaptive management program could reduce critical uncertainties, inform lake regulation schedule planning, and enhance real-time lake operations. For example, moving from quarterly transect sampling of SAV to more frequent sampling at just a few representative sites might provide more actionable information and lead to a better understanding of the effects of antecedent conditions. Further, the Lake Okeechobee Environment Model is a tool to use in concert with regional hydrologic and ecological models to evaluate the implications of alternative regulation schedules and lake operations, particularly as new data become available to refine the model's SAV component.

A CERP MID-COURSE ASSESSMENT:
SOUND DECISION MAKING FOR THE FUTURE

The Everglades of 2050 and beyond will differ from what was originally envisioned when the CERP was developed. The original CERP plan was formulated based on a pre-drainage vision of the historical Everglades and the assumption that specific rainfall and temperature time series observed in the past captured the full range of variability expected throughout the 21st century. There is now ample evidence that the South Florida climate is changing. There is general consensus that temperatures will increase over time, although considerable uncertainty about future rainfall patterns remains. There is also compelling recent evidence that sea-level rise is accelerating. These changes will have profound impacts on the South Florida ecosystem and the related challenges of providing flood protection and meeting future water and recreational demands.

CERP agencies should conduct a mid-course assessment that rigorously considers the future of the South Florida ecosystem. New information about climate variability, climate change, and sea-level rise in South Florida continues to emerge, and many of these changes will impact the capacity for the CERP to meet its goals. Although the SFWMD has begun to conduct these types of analyses for planning and management projects outside of the restoration, CERP agencies do not adequately account for these changes when planning projects, and they have not systematically analyzed these threats in the context of the CERP. Restoration is likely to create important benefits that increase the resilience of the ecosystem in the face of climate change, but these benefits have not been studied or quantified. A systemwide, program-level analysis should assess the resilience and robustness of the CERP to the changing conditions that will drive the Everglades of the future. A mid-course assessment should include systemwide modeling of interactions among both authorized and planned projects under scenarios of future possible climate and sea level–rise conditions. This assessment is essential to communicate the benefits of the CERP to stakeholders, guide project sequencing and investment decisions, and manage the restoration under changing conditions. Now that several major project planning efforts are nearing completion and the vision for CERP storage is largely developed, which will require decades to construct at current funding levels, the time is right for a mid-course assessment.

A science program focused on understanding the impacts of current and future stressors on the South Florida ecosystem is needed to ensure that CERP agencies have the latest scientific information and tools to successfully plan and implement the restoration program. This report has highlighted the ongoing research advances and science that are needed to address issues of vital importance for the long-term success of restoration investments, such as understanding peat collapse, saltwater intrusion, and the management of invasive species. Ensuring that investigative research and advances in tools and understanding are useful in a policy context requires a programmatic approach directly linked to the CERP effort, which may be best championed by an independent Everglades Lead Scientist empowered to coordinate and promote needed scientific advances.

1

Introduction

The Florida Everglades, formerly a large and diverse aquatic ecosystem, has been dramatically altered during the past century by an extensive water control infrastructure designed to increase regional economic productivity through improved flood management, urban water supply, and agricultural production (Davis and Ogden, 1994). Shaped by the slow flow of water, its vast terrain of sawgrass plains, ridges, sloughs, and tree islands supported a high diversity of plant and animal habitats. This natural landscape also served as a sanctuary for Native Americans. However, large-scale changes to the landscape have diminished the natural resources, and by the mid- to late-20th century many of the area's defining natural characteristics had been lost. The remnants of the original Everglades (see Figure 1-1 and Box 1-1) now compete for vital water with urban and agricultural interests, and contaminated runoff from these two activities impairs the South Florida ecosystem.

Recognition of past declines in environmental quality, combined with continuing threats to the natural character of the remaining Everglades, led to initiation of large-scale restoration planning in the 1990s and the launch of the Comprehensive Everglades Restoration Plan (CERP) in 2000. This unprecedented project envisioned the expenditure of billions of dollars in a multidecadal effort to achieve ecological restoration by reestablishing the hydrologic characteristics of the Everglades, where feasible, and to create a water system that simultaneously serves the needs of both the natural and the human systems of South Florida. Within the social, economic, and political latticework of the 21st century, restoration of the South Florida ecosystem is now under way and represents one of the most ambitious ecosystem renewal projects ever conceived. This report represents the seventh independent assessment of the CERP's progress by the Committee on Independent Scientific Review of Everglades Restoration Progress (CISRERP) of the National Academies of Sciences, Engineering, and Medicine.

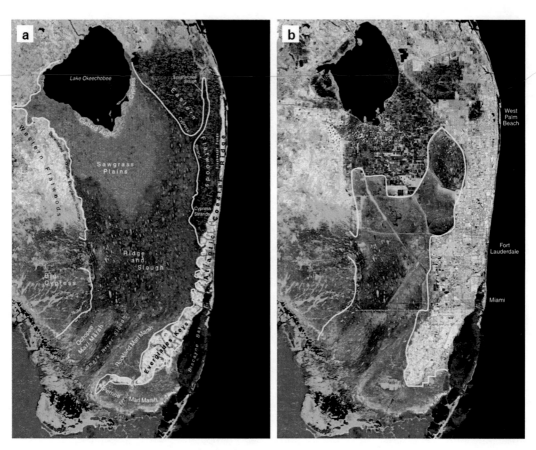

FIGURE 1-1 Reconstructed (a) pre-drainage (circa 1850) and (b) current (1994) satellite images of the Everglades ecosystem.

NOTE: The yellow line in (a) outlines the historical Everglades ecosystem, and the yellow line in (b) outlines the remnant Everglades ecosystem as of 1994.

SOURCE: Courtesy of C. McVoy, J. Obeysekera, and W. Said, South Florida Water Management District.

THE NATIONAL ACADEMIES AND EVERGLADES RESTORATION

The National Academies has provided scientific and technical advice related to the Everglades restoration since 1999. The National Academies' Committee on Restoration of the Greater Everglades Ecosystem (CROGEE), which operated from 1999 until 2004, was formed at the request of the South Florida Ecosystem

BOX 1-1
Geographic Terms

This box defines some key geographic terms used throughout this report.

- The **Everglades,** the **Everglades ecosystem,** or the **remnant Everglades ecosystem** refers to the present areas of sawgrass, marl prairie, and other wetlands and estuaries south of Lake Okeechobee (Figure 1-1b).
- The **original, historical,** or **pre-drainage Everglades** refers to the areas of sawgrass, marl prairie, and other wetlands and estuaries south of Lake Okeechobee that existed prior to the construction of drainage canals beginning in the late 1800s (Figure 1-1a).
- The **Everglades watershed** is the drainage that encompasses the Everglades ecosystem but also includes the Kissimmee River watershed and other smaller watersheds north of Lake Okeechobee that ultimately supply water to the Everglades ecosystem.
- The **South Florida ecosystem** (also known as the Greater Everglades Ecosystem; see Figure 1-2) extends from the headwaters of the Kissimmee River near Orlando through Lake Okeechobee and the Everglades into Florida Bay and ultimately the Florida Keys. The boundaries of the South Florida ecosystem are determined by the boundaries of the South Florida Water Management District, the southernmost of the state's five water management districts, although they approximately delineate the boundaries of the South Florida watershed. This designation is important and helpful to the restoration effort because, as many publications have made clear, taking a watershed approach to ecosystem restoration is likely to improve the results, especially when the ecosystem under consideration is as water dependent as the Everglades (NRC, 1999, 2004).
- The **Water Conservation Areas** (WCAs) include WCA-1 (the Arthur R. Marshall Loxahatchee National Wildlife Refuge), WCA-2A, -2B, -3A, and -3B (see Figure 1-2).

The following represent legally defined geographic terms used in this report:

- The **Everglades Protection Area** is defined in the Everglades Forever Act as comprising WCA-1, -2A, -2B, -3A, and -3B and Everglades National Park.
- The **natural system** is legally defined in the Water Resources Development Act of 2000 (WRDA 2000) as all land and water managed by the federal government or the state within the South Florida ecosystem (see Figure 1-3). "The term 'natural system' includes (i) water conservation areas; (ii) sovereign submerged land; (iii) Everglades National Park; (iv) Biscayne National Park; (v) Big Cypress National Preserve; (vi) other Federal or State (including a political subdivision of a State) land that is designated and managed for conservation purposes; and (vii) any tribal land that is designated and managed for conservation purposes, as approved by the tribe" (WRDA 2000).

Many maps in this report include shorthand designations that use letters and numbers for engineered additions to the South Florida ecosystem. For example, canals are labeled C-#; levees and associated borrow canals as L-#; and structures, such as culverts, locks, pumps, spillways, control gates, and weirs, as S-# or G-#.

FIGURE 1-2 The South Florida ecosystem.

SOURCE: © International Mapping Associates.

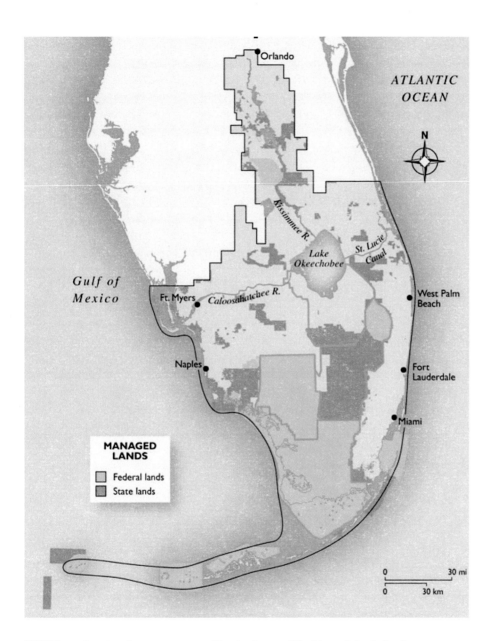

FIGURE 1-3 Land and waters managed by the State of Florida and the federal government as of December 2005 for conservation purposes within the South Florida ecosystem.

SOURCE: Based on data compiled by Florida State University's Florida Natural Areas Inventory (http://www.fnai.org/gisdata.cfm). © International Mapping Associates.

Restoration Task Force (hereafter, simply the Task Force), an intergovernmental body established to facilitate coordination in the restoration effort, and the committee produced six reports (NRC, 2001, 2002a,b, 2003a,b, 2005). The National Academies' Panel to Review the Critical Ecosystem Studies Initiative produced an additional report in 2003 (NRC, 2003c; see Appendix A). The Water Resources Development Act of 2000 (WRDA 2000) mandated that the U.S. Department of the Army, the Department of the Interior, and the State of Florida, in consultation with the Task Force, establish an independent scientific review panel to evaluate progress toward achieving the natural system restoration goals of the CERP. The National Academies' CISRERP was therefore established in 2004 under contract with the U.S. Army Corps of Engineers. After publication of each of the first six biennial reviews (NASEM, 2016; NRC, 2007, 2008, 2010, 2012, 2014; see Appendix A for the report summaries), some members rotated off the committee and some new members were added.

The committee is charged to submit biennial reports that address the following items:

1. An assessment of progress in restoring the natural system, which is defined by section 601(a) of WRDA 2000 as all the land and water managed by the federal government and state within the South Florida ecosystem (see Figure 1-3 and Box 1-1);
2. A discussion of significant accomplishments of the restoration;
3. A discussion and evaluation of specific scientific and engineering issues that may impact progress in achieving the natural system restoration goals of the plan; and
4. An independent review of monitoring and assessment protocols to be used for evaluation of CERP progress (e.g., CERP performance measures, annual assessment reports, assessment strategies).

Given the broad charge, the complexity of the restoration, and the continually evolving circumstances, the committee did not presume it could cover all issues that affect restoration progress in any single report. This report builds on the past reports by this committee (NASEM, 2016; NRC, 2007, 2008, 2010, 2012, 2014) and emphasizes restoration progress since 2016, high-priority scientific and engineering issues that the committee judged to be relevant to this time frame, and other issues that have impacted the pace of progress. The committee focused particularly on issues for which the "timing was right"—that is, where the committee's advice could be useful relative to the decision-making time frames—and on topics that had not been fully addressed in past National Academies Everglades reports. Interested readers should look to past reports by this committee to find detailed discussions of important topics, such as new information impacting the CERP (NASEM, 2016), climate change (NASEM, 2016; NRC, 2014), invasive species (NRC, 2014), science synthesis (NRC, 2012), the human context for the CERP (NRC, 2010), economic valuation of ecosystem services (NRC, 2010), water quality and quantity challenges and trajectories (NRC, 2010, 2012), Modified Water Deliveries to Everglades National Park (NRC, 2008), Lake Okeechobee (NRC, 2008), and incremental adaptive restoration (NRC, 2007). Past reports have also discussed various aspects of the CERP monitoring and assessment plan (NRC, 2004, 2008, 2010, 2012, 2014).

The committee met in person five times during the course of this review; received briefings at its public meetings from agencies, organizations, and individuals involved in the restoration, as well as from the public; and took several field trips to sites with restoration activities (see Acknowledgments). In addition to information received during the meetings, the committee based its assessment of progress on information in relevant CERP and non-CERP restoration documents. The committee's conclusions and recommendations were also informed by a review of relevant scientific literature and the experience and knowledge of the committee members in their fields of expertise. The committee was unable to consider in any detail new materials received after May 2018.

REPORT ORGANIZATION

In Chapter 2, the committee provides an overview of the CERP in the context of other ongoing restoration activities and discusses the restoration goals that guide the overall effort.

In Chapter 3, the committee analyzes the natural system restoration progress associated with CERP and non-CERP projects, along with programmatic factors and planning efforts that affect future progress.

In Chapter 4, the committee performs an in-depth review of CERP monitoring, with particular emphasis on project-level monitoring and assessment.

In Chapter 5, the committee synthesizes recent information on Lake Okeechobee and the effect of water levels on lake ecology to inform forward-looking systemwide planning and operations decisions.

In Chapter 6, the committee discusses the value of a mid-course assessment of the CERP and research on future stressors to support restoration decision making.

2

The Restoration Plan in Context

This chapter sets the stage for the seventh of this committee's biennial assessments of restoration progress in the South Florida ecosystem. Background for understanding the project is provided through descriptions of the ecosystem decline, restoration goals, the needs of a restored ecosystem, and the specific activities of the restoration project.

BACKGROUND

The Everglades once encompassed about 3 million acres of slow-moving water and associated biota that stretched from Lake Okeechobee in the north to the Florida Keys in the south (Figures 1-1a and 2-1a). The conversion of the Everglades wilderness into an area of high agricultural productivity and cities was a dream of 19th-century investors, and projects begun between 1881 and 1894 affected the flow of water in the watershed north of Lake Okeechobee. These early projects included dredging canals in the Kissimmee River Basin and constructing a channel connecting Lake Okeechobee to the Caloosahatchee River and, ultimately, the Gulf of Mexico. By the late 1800s, more than 50,000 acres north and west of the lake had been drained and cleared for agriculture (Grunwald, 2006). In 1907, Governor Napoleon Bonaparte Broward created the Everglades Drainage District to construct a vast array of ditches, canals, dikes, and "improved" channels. By the 1930s, Lake Okeechobee had a second outlet, through the St. Lucie Canal, leading to the Atlantic Ocean, and 440 miles of other canals altered the hydrology of the Everglades (Blake, 1980). After hurricanes in 1926 and 1928 resulted in disastrous flooding from Lake Okeechobee, the U.S. Army Corps of Engineers (USACE) replaced the small berm that bordered the southern edge of the lake with the massive Herbert Hoover Dike, which was eventually expanded in the 1960s to encircle the lake. The hydrologic end product of these drainage activities was the drastic reduction of natural water

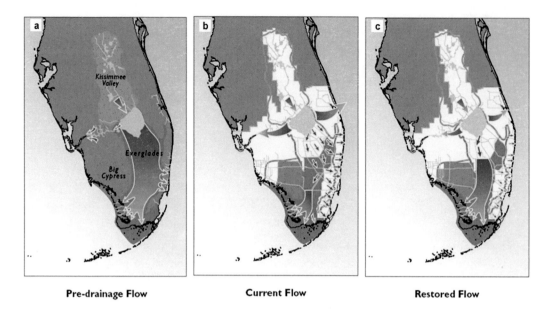

Pre-drainage Flow   **Current Flow**   **Restored Flow**

FIGURE 2-1 Water flow in the Everglades under (a) historical conditions, (b) current conditions, and (c) conditions envisioned upon completion of the Comprehensive Everglades Restoration Plan.

SOURCE: Graphics provided by USACE, Jacksonville District.

---

storage within the system and an increased susceptibility to drought and desiccation in the southern reaches of the Everglades (NRC, 2005).

After further flooding in 1947 and increasing demands for improved agricultural production and flood management for the expanding population centers on the southeast Florida coast, the U.S. Congress authorized the Central and Southern Florida (C&SF) Project. This project provided flood management and urban and agricultural water supply by straightening 103 miles of the meandering Kissimmee River, expanding the Herbert Hoover Dike, constructing a levee along the eastern boundary of the Everglades to prevent flows into the southeastern urban areas, establishing the 700,000-acre Everglades Agricultural Area (EAA) south of Lake Okeechobee, and creating a series of Water Conservation Areas (WCAs) in the remaining space between the lake and Everglades National Park (Light and Dineen, 1994). The eastern levee isolated about 100,000 acres of the Everglades ecosystem, making it available for development (Lord, 1993). In total, urban and agricultural development have reduced the Everglades to about one-half its pre-drainage area (see Figure 1-1b; Davis and Ogden, 1994) and have contaminated its waters with chemicals such as phosphorus, nitrogen, sulfur, mercury, and pesticides. Associated drainage and flood management structures, including the C&SF Project, have diverted large quantities of water directly east and west to the northern estuaries, thereby reducing the dominantly southward freshwater flows and natural water storage that defined the ecosystem (see Figure 2-1b).

The profound hydrologic alterations were accompanied by many changes in the biotic communities in the ecosystem, including reductions and changes in the composition, distribution, and abundance of the populations of wading birds. Today, the federal government has listed 78 plant and animal species in South Florida as threatened or endangered, with many more included on state lists. Some distinctive Everglades habitats, such as custard apple forests and peripheral wet prairie, have disappeared altogether, while other habitats are severely reduced in area (Davis and Ogden, 1994; Marshall et al., 2004). Approximately 1 million acres are contaminated with mercury from atmospheric deposition (McPherson and Halley, 1996; Orem et al., 2011). Phosphorus from agricultural runoff has impacted water quality in large portions of the Everglades and has been particularly problematic in Lake Okeechobee (Flaig and Reddy, 1995). The Caloosahatchee and St. Lucie estuaries, including parts of the Indian River Lagoon, have been greatly altered by high and extremely variable freshwater discharges that bring nutrients and contaminants and disrupt salinity regimes (Doering, 1996; Doering and Chamberlain, 1999).

At least as early as the 1920s, private citizens were calling attention to the degradation of the Florida Everglades (Blake, 1980). However, by the time Marjory Stoneman Douglas's classic book *The Everglades: River of Grass* was published in 1947 (the same year that Everglades National Park was dedicated), the South Florida ecosystem had already been altered extensively. Beginning in the 1970s, prompted by concerns about deteriorating conditions in Everglades National Park and other parts of the South Florida ecosystem, the public, as well as the federal and state governments, directed increased attention to the adverse ecological effects of the flood management and irrigation projects (Kiker et al., 2001; Perry, 2004). By the late 1980s it was clear that various minor corrective measures undertaken to remedy the situation were insufficient. As a result, a powerful political consensus developed among federal agencies, state agencies and commissions, Native American tribes, county governments, and conservation organizations that a large restoration effort was needed in the Everglades (Kiker et al., 2001). This recognition culminated in the Comprehensive Everglades Restoration Plan (CERP), authorized by Congress in 2000, which builds on other ongoing restoration activities of the state and federal governments to create what was at the time the most ambitious restoration effort in the nation's history.

RESTORATION GOALS FOR THE EVERGLADES

Several goals have been articulated for the restoration of the South Florida ecosystem, reflecting the various restoration programs. The South Florida Ecosystem Restoration Task Force (hereafter, simply the Task Force), an intergovernmental body established to facilitate coordination in the restoration effort, has three broad strategic goals: (1) "get the water right," (2) "restore, preserve, and protect natural habitats and species," and (3) "foster compatibility of the built and natural systems" (SFERTF, 2000). These goals encompass, but are not limited to, the CERP. The Task Force works to coordinate and build consensus among the many non-CERP restoration initiatives that support these broad goals.

The goal of the CERP, as stated in the Water Resources Development Act of 2000 (WRDA 2000), is "restoration, preservation, and protection of the South Florida Ecosystem while providing for other water-related needs of the region, including water supply and flood protection." The Programmatic Regulations (33 CFR § 385.3) that guide implementation of the CERP further clarify this goal by defining restoration as "the recovery and protection of the South Florida ecosystem so that it once again achieves and sustains those essential hydrological and biological characteristics that defined the undisturbed South Florida ecosystem." These defining characteristics include a large areal extent of interconnected wetlands, extremely low concentrations of nutrients in freshwater wetlands, sheet flow, healthy and productive estuaries, resilient plant communities, and an abundance of native wetland animals (DOI and USACE, 2005). Although development has permanently reduced the areal extent of the Everglades ecosystem, the CERP hopes to recover many of the Everglades' original characteristics and natural ecosystem processes in the remnant Everglades. At the same time, the CERP is charged to maintain levels of flood protection (as of 2000) and was designed to provide for other water-related needs, including water supply (DOI and USACE, 2005).

Although the CERP contributes to each of the Task Force's three goals, it focuses primarily on restoring the hydrologic features of the undeveloped wetlands remaining in the South Florida ecosystem, on the assumption that improvements in ecological conditions will follow. Originally, "getting the water right" had four components—quality, quantity, timing, and distribution. However, the hydrologic properties of flow, encompassing the concepts of direction, velocity, and discharge, have been recognized as an important component of getting the water right that had previously been overlooked (NRC, 2003c; SCT, 2003). Numerous studies have supported the general approach to getting the water right (Davis and Ogden, 1994; NRC, 2005; SSG, 1993), although it is widely recognized that recovery of the native habitats and species in South Florida may require restoration efforts in addition to getting the water right, such as controlling non-native species and reversing the decline in the spatial extent and compartmentalization of the natural landscape (SFERTF, 2000; SSG, 1993).

The goal of ecosystem restoration can seldom be the exact re-creation of some historical or preexisting state because physical conditions, driving forces, and boundary conditions usually have changed and are not fully recoverable. Rather, restoration is better viewed as the process of assisting the recovery of a degraded ecosystem to the point where it contains sufficient biotic and abiotic resources to continue its functions without further assistance in the form of energy or other resources from humans (NRC, 1996; Society for Ecological Restoration International Science & Policy Working Group, 2004). The term *ecosystem rehabilitation* may be more appropriate when the objective is to improve conditions in a part of the South Florida ecosystem to at least some minimally acceptable level that allows the restoration of the larger ecosystem to advance. However, flood management remains a critical aspect of the CERP design because improving hydrology and sheet flow in extensive wetland areas has the potential, through seepage, to flood adjacent urban and agricultural areas. Artificial storage will be required to replace the lost natural storage in the system (NRC, 2005), and groundwater management also requires attention to boundaries between developed and natural areas. For these and other reasons, even when the CERP is complete, it will require large inputs of energy and human effort to operate and maintain pumps, stormwater treatment areas, canals and levees, and reservoirs, and to continue to manage non-native species. Thus, for the foreseeable future, the CERP does not envision ecosystem restoration or rehabilitation that returns the ecosystem to a state where it can "manage itself."

The broad CERP goals should be interpreted in the context of the complex Everglades ecosystem in order to guide restoration efforts. Early restoration was motivated by ambitious albeit generalized expectations for the ecosystem. For example, the CERP conceptual plan, also called the Yellow Book (USACE and SFWMD, 1999), stated: "At all levels in the aquatic food chains, the numbers of such animals as crayfish, minnows, sunfish, frogs, alligators, herons, ibis, and otters, will markedly increase." Currently the goals for the restoration upon which policy makers agree (USACE et al., 2007) are largely qualitative, indicating a desired direction of change for a number of indicators, without a quantitative objective, providing no clear expectation of how the success of restoration efforts should collectively be assessed. Continued investment in Everglades restoration proceeds based on improving the current undesirable state of the system rather than toward a specific set of quantitative characteristics desired for the future South Florida ecosystem.

An additional factor challenging the ability of the restoration efforts to meet the "essential hydrological and biological characteristics that defined the undisturbed South Florida ecosystem" is ongoing climate change, including changes in precipitation patterns, sea-level rise, and ocean warming. Not only did irreversible changes occur since the 19th century, but also, since the development of the CERP, sea levels have risen approximately 6-7 cm and future projections call for further increases of as much as 2 meters in South Florida in the 21st century (NOAA, 2017).

Implicit in the understanding of ecosystem restoration is the recognition that natural systems are self-designing and dynamic, and therefore it is not possible to know in advance exactly what can or will be achieved. Thus, ecosystem restoration proceeds in the face of scientific uncertainty and must consider a range of possible future conditions. NASEM (2016) discusses the challenges to restoration goals arising from major changes that have occurred since the inception of the CERP in 1999, and further consideration of these issues is provided in this report.

What Restoration Requires

Restoring the South Florida ecosystem to a desired ecological landscape requires reestablishment of critical processes that sustain its functioning. Although getting the water right is the oft-stated and immediate goal, the restoration ultimately aims to restore the distinctive characteristics of the historical ecosystem to the remnant Everglades (DOI and USACE, 2005). Getting the water right is a means to that end, not the end itself. The hydrologic and ecologic characteristics of the historical Everglades serve as general restoration goals for a functional (albeit reduced in size) Everglades ecosystem. The first Committee on Independent Scientific Review of Everglades Restoration Progress identified five critical components of Everglades restoration (NRC, 2007):

1. Enough water storage capacity combined with operations that allow for appropriate volumes of water to support healthy estuaries and the return of sheet flow through the Everglades ecosystem while meeting other demands for water;

2. Mechanisms for delivering and distributing the water to the natural system in a way that resembles historical flow patterns, affecting volume, depth, velocity, direction, distribution, and timing of flows;

3. Barriers to eastward seepage of water so that higher water levels can be maintained in parts of the Everglades ecosystem without compromising the current levels of flood protection of developed areas as required by the CERP;

4. Methods for securing water quality conditions compatible with restoration goals for a natural system that was inherently extremely nutrient poor, particularly with respect to phosphorus; and

5. Retention, improvement, and expansion of the full range of habitats by

preventing further losses of critical wetland and estuarine habitats, and by protecting lands that could usefully be part of the restored ecosystem.

If these five critical components of restoration are achieved and the difficult problem of invasive species can be managed, then the basic physical, chemical, and biological processes that created the historical Everglades can once again work to create and sustain a functional mosaic of biotic communities that resemble what was distinctive about the historical Everglades albeit of a reduced scale.

The history of the Everglades and ongoing global climate change will make replication of the pre-drainage system impossible. Because of the historical changes that have occurred through engineered structures, urban development, introduced species, and other factors, the paths taken by the ecosystem and its components in response to restoration efforts will not retrace the paths taken to reach current conditions. End results will also often differ from the historical system as climate change and sea-level rise, permanently established invasive species, and other factors have moved the ecosystem away from its historical state (Hiers et al., 2012) and will continue to change the restored system in the future. The specific nature and extent of the functional mosaic thus depends on not only the degree to which the five critical components can be achieved but also future precipitation patterns, rising sea levels, marine incursion into estuaries and coastal wetlands, as well as continued investment in water and ecological management.

Even if the restored system does not exactly replicate the historical system, or reach all the biological, chemical, and physical targets, the reestablishment of natural processes and dynamics should result in a viable and valuable Everglades ecosystem under current conditions. The central principle of ecosystem management is to provide for the natural processes that historically shaped an ecosystem, because ecosystems are characterized by the processes that regulate them. How the reestablished processes interact with future changes within and external to the system will determine the future character of the ecosystem, its species, and communities.

RESTORATION ACTIVITIES

Several restoration programs, including the largest of the initiatives, the CERP, are now under way. The CERP often builds upon non-CERP activities (also called "foundation projects"), many of which are essential to the effectiveness of the CERP. The following section provides a brief overview of the CERP and some of the major non-CERP activities.

Comprehensive Everglades Restoration Plan

WRDA 2000 authorized the CERP as the framework for modifying the C&SF Project. Considered a blueprint for the restoration of the South Florida ecosystem, the CERP is led by two organizations with considerable expertise managing the water resources of South Florida—the USACE, which built most of the canals and levees throughout the region, and the South Florida Water Management District (SFWMD), the state agency with primary responsibility for operating and maintaining this complicated water collection and distribution system.

The CERP conceptual plan (USACE and SFWMD, 1999) proposes major alterations to the C&SF Project in an effort to reverse decades of ecosystem decline. The Yellow Book includes approximately 50 major projects consisting of 68 project components to be constructed at a cost of approximately $16.4 billion (estimated in 2014 dollars, including program coordination and monitoring costs; USACE and DOI, 2016; Figure 2-2). Major components of the restoration plan focus on restoring the quantity, quality, timing, and distribution of water for the South Florida ecosystem. The Yellow Book outlines the major CERP components, including the following:

- **Conventional surface-water storage reservoirs.** The Yellow Book includes plans for approximately 1.5 million acre-feet (AF) of storage, located north of Lake Okeechobee, in the St. Lucie and Caloosahatchee basins, in the EAA, and in Palm Beach, Broward, and Miami-Dade counties.
- **Aquifer storage and recovery (ASR).** The Yellow Book proposes to provide substantial water storage through ASR, a highly engineered approach that would use a large number of wells built around Lake Okeechobee, in Palm Beach County, and in the Caloosahatchee Basin to store water approximately 1,000 feet below ground.
- **In-ground reservoirs.** The Yellow Book proposes additional water storage in quarries created by rock mining.
- **Stormwater treatment areas (STAs).** The CERP contains plans for additional constructed wetlands that will treat agricultural and urban runoff water before it enters natural wetlands.[1]

---

[1] Although some STAs are included among CERP projects, the USACE has clarified its policy on federal cost sharing for water quality features. A memo from the Assistant Secretary of the Army (Civil Works) (USACE, 2007a) states: "Before there can be a Federal interest to cost share a WQ [water quality] improvement feature, the State must be in compliance with WQ standards for the current use of the water to be affected and the work proposed must be deemed essential to the Everglades restoration effort." The memo goes on to state, "the Yellow Book specifically envisioned that the State would be responsible for meeting water quality standards." However, the Secretary of the Army can recommend to Congress that projects features deemed "essential to Everglades restoration" be cost shared. In such cases, the state is responsible for 100 percent of the costs to treat water to

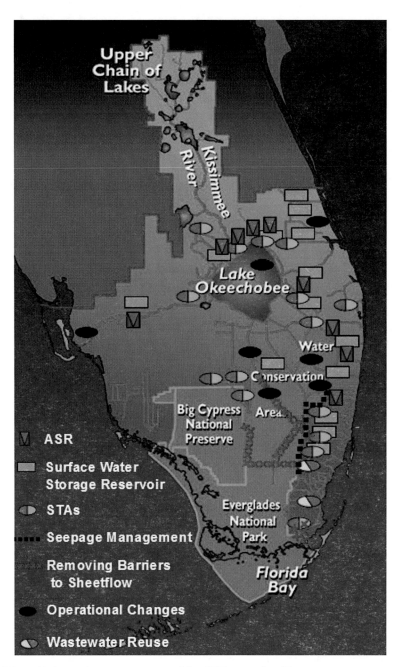

FIGURE 2-2 Major project components of the CERP as outlined in 1999.

SOURCE: Courtesy of Laura Mahoney, USACE.

- **Seepage management.** The Yellow Book outlines seepage management projects to prevent unwanted loss of water from the remnant Everglades through levees and groundwater flow. The approaches include adding impermeable barriers to the levees, installing pumps near levees to redirect lost water back into the Everglades, and holding water levels higher in undeveloped areas between the Everglades and the developed lands to the east.
- **Removing barriers to sheet flow.** The CERP includes plans for removing 240 miles of levees and canals, to reestablish shallow sheet flow of water through the Everglades ecosystem.
- **Rainfall-driven water management.** The Yellow Book includes operational changes in the water delivery schedules to the WCAs and Everglades National Park to mimic more natural patterns of water delivery and flow through the system.
- **Water reuse and conservation.** To address shortfalls in water supply, the Yellow Book proposes two advanced wastewater treatment plants so that the reclaimed water could be discharged to wetlands along Biscayne Bay or used to recharge the Biscayne aquifer.

The largest portion of the budget is devoted to storage projects and to acquiring the lands needed for them.

The modifications to the C&SF Project embodied in the CERP were originally expected to take more than three decades to complete (and will likely now take much longer), and to be effective they require a clear strategy for managing and coordinating restoration efforts. The Everglades Programmatic Regulations (33 CFR Part 385) state that decisions on CERP implementation are made by the USACE and the SFWMD (or any other local project sponsors), in consultation with the Department of the Interior, the Environmental Protection Agency (EPA), the Department of Commerce, the Miccosukee Tribe of Indians of Florida, the Seminole Tribe of Florida, the Florida Department of Environmental Protection, and other federal, state, and local agencies (33 CFR Part 385).

WRDA 2000 endorses the use of an adaptive management framework for the restoration process, and the Programmatic Regulations (33 CFR §385.31[a]) formally establish an adaptive management program that will

> assess responses of the South Florida ecosystem to implementation of the Plan; . . . [and] seek continuous improvement of the Plan based upon new information resulting from changed or unforeseen circumstances, new scientific and technical information, new or updated modeling; information developed

---

state standards for its current use, and federal cost sharing is determined based on the additional treatment needed to meet the requirements of Everglades restoration (K. Taplin, USACE, personal communication, 2018).

through the assessment principles contained in the Plan; and future authorized changes to the Plan. . . .

An interagency body called Restoration, Coordination, and Verification (RECOVER) was established early in the development of the CERP to ensure that sound science is used in the restoration. The RECOVER leadership group oversees the monitoring and assessment program that will evaluate the progress of the CERP toward restoring the natural system and will assess the need for changes to the plan through the adaptive management process (see also Chapter 5).

Non-CERP Restoration Activities

When Congress authorized the CERP in WRDA 2000, the SFWMD, the USACE, the National Park Service, and the U.S. Fish and Wildlife Service were already implementing several activities intended to restore key aspects of the Everglades ecosystem. These non-CERP initiatives are critical to the overall restoration progress. In fact, the CERP's effectiveness was predicated upon the completion of many of these projects, which include Modified Water Deliveries to Everglades National Park (Mod Waters), C-111 South Dade, and state water quality treatment projects (see Figure 2-3). Several additional projects are also under way to meet the broad restoration goals for the South Florida ecosystem and associated legislative mandates. They include extensive water quality treatment initiatives and programs to establish best management practices (BMPs) to reduce nutrient loading. Recent progress on key non-CERP projects with critical linkages to the CERP are described in Chapter 3.

Major Developments and Changing Context Since 2000

Several major program-level developments have occurred since the CERP was launched that have affected the pace and focus of CERP efforts. In 2004, Florida launched Acceler8, a plan to hasten the pace of project implementation that was bogged down by the slow federal planning process (for further discussion of Acceler8, see NRC, 2007). Acceler8 originally included 11 CERP project components and 1 non-CERP project, and although the state was unable to complete all the original tasks, the program led to increased state investment and expedited project construction timelines for several CERP projects.

Operation of Lake Okeechobee has been modified twice since the CERP was developed in ways that have reduced total storage. In April 2000, the Water Supply and Environment (WSE) regulation schedule was implemented to reduce high-water impacts on the lake's littoral zone and to reduce harmful high discharges to the St. Lucie and Caloosahatchee estuaries. The regulation

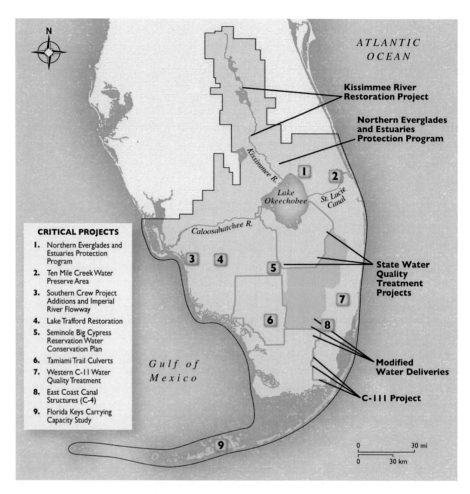

FIGURE 2-3 Locations of major non-CERP initiatives.

SOURCE: © International Mapping Associates.

schedule was changed again in 2008 to reduce the risk of failure of the Herbert Hoover Dike until the USACE could make critical repairs. This resulted in a loss of 564,000 AF of potential storage from the regional system (see NASEM, 2016).

In the years since the CERP was launched, the state of Florida has increasingly encouraged the use of alternative water supplies—including wastewater, stormwater, and excess surface water—to meet future water demands (e.g., FDEP, 2015). In 2006, the SFWMD passed the Lower East Coast Regional Water

Availability Rule, which caps groundwater withdrawals at 2006 levels, requiring urban areas to meet increased demand through a combination of conservation and alternative water supplies. In 2007, the Florida legislature mandated that ocean wastewater discharges in South Florida be eliminated and 60 percent of those discharges be reused by 2025 (Section 403.086[9], Florida Statute), representing approximately 180 million gallons per day of new water supply for the Lower East Coast. It remains unclear whether or how these new initiatives and mandates will affect the expectations for agricultural and urban water supply from the CERP, particularly because the capture of excess surface water is a key element of the CERP.

In 2010, EPA issued its court-ordered Amended Determination, which directed the state of Florida to correct deficiencies in meeting the narrative and numeric nutrient criteria in the Everglades Protection Area (EPA, 2010). In 2012, the state of Florida launched its Restoration Strategies Regional Water Quality Plan, which was approved by EPA and the court as an alternative means to address the Amended Determination. The state of Florida is currently in the process of constructing approximately 6,500 acres of new STAs and three flow equalization basins (116,000 AF; see Chapter 3). These water quality treatment improvements are designed so that water leaving the STAs will meet a new water quality–based effluent limit (WQBEL) to comply with the 10 parts per billion (ppb) total phosphorus water quality criterion for the Everglades Protection Area.[2]

Changing Understanding of Restoration Challenges

Much new knowledge has been gained since the launch of the CERP that provides a new understanding of restoration challenges and opportunities and informs future restoration planning and management. Considering the many advances in knowledge since 1999, climate change and sea-level rise are among the most significant. As outlined in NASEM (2016), changes in precipitation and evapotranspiration are expected to have substantial impacts on CERP outcomes. Downscaled precipitation projections remain uncertain and range from modest increases to sizable decreases for South Florida, and research continues locally and nationally to improve these projections. Sea-level rise is already affecting the distribution of Everglades habitats and causing coastal flooding in some

---

[2] The WQBEL is a numeric discharge limit used to regulate permitted discharges from the STAs so as to not exceed a long-term geometric mean of 10 ppb within the Everglades Protection Area. This numeric value is now translated into a flow-weighted mean (FWM) total phosphorus (TP) concentration and applied to each STA discharge point, which now must meet the following: (1) the STAs are in compliance with WQBEL when the TP concentration of STA discharge point does not exceed an annual FWM of 13 ppb in more than 3 out of 5 years, and (2) annual FWM of 19 ppb in any water year (Fla. Stat. §373.4592; EPA, 2010; Leeds, 2014).

low-lying urban areas (see Chapter 6). CERP planners are now evaluating all future restoration benefits in the context of low, medium, and high sea-level rise projections, although NRC (2014) noted the need for greater consideration of climate change and sea-level rise in CERP project and program planning.

Since the CERP was developed, the significance of invasive species management on the success of restoration has also been recognized by the South Florida Ecosystem Restoration Task Force and its member agencies.[3] Non-native species constitute a substantial proportion of the current biota of the Everglades. The approximately 250 non-native plants species are about 16 percent of the regional flora (see NRC, 2014). South Florida has a subtropical climate with habitats that are similar to those from which many of the invaders originate, with relatively few native species in many taxa to compete with introduced ones. Some species, especially of introduced vascular plants and reptiles, have had dramatic effects on the structure and functioning of Everglades ecosystems, and necessitate aggressive management and early detection of new high-risk invaders to ensure that ongoing CERP efforts to get the water right allow native species to prosper instead of simply enhancing conditions for invasive species.

SUMMARY

The Everglades ecosystem is one of the world's ecological treasures, but for more than a century the installation of an extensive water management infrastructure has changed the geography of South Florida and has facilitated extensive agricultural and urban development. These changes have had profound ancillary effects on regional hydrology, vegetation, and wildlife populations. The CERP, a joint effort led by the state and federal governments and launched in 2000, seeks to reverse the general decline of the ecosystem. Since 2000, the legal context for the CERP and other major Everglades restoration efforts has evolved and the scientific understanding of Everglades restoration and its current and future stressors has expanded, and the programs continue to adapt. Implementation progress is discussed in detail in Chapter 3.

---

[3] See http://www.evergladesrestoration.gov/content/ies/.

3

Restoration Progress

This committee is charged with the task of discussing accomplishments of the restoration and assessing "the progress toward achieving the natural system restoration goals of the Comprehensive Everglades Restoration Plan [CERP]" (see Chapter 1 for the statement of task and Chapter 2 for a discussion of restoration goals). In this chapter, the committee updates the National Academies' previous assessments of CERP and related non-CERP restoration projects (NASEM, 2016; NRC, 2007, 2008, 2010, 2012, 2014). This chapter also addresses programmatic and implementation progress and discusses the ecosystem benefits resulting from the progress to date.

PROGRAMMATIC PROGRESS

To assess programmatic progress the committee reviewed a set of primary issues that influence CERP progress toward its overall goals of ecosystem restoration. These issues, described in the following sections, relate to project authorization, funding, and project scheduling.

Project Authorization

Once project planning is complete, CERP projects with costs exceeding $25 million must be individually authorized by Congress. Water Resources Development Acts (WRDAs) have served as the mechanism to congressionally authorize U.S. Army Corps of Engineers (USACE) projects, and the CERP planning process was developed with the assumption that WRDAs would be passed every 2 years. This, however, has not occurred. In the 18 years since the CERP was launched in WRDA 2000, three WRDA bills have been enacted:

- WRDA 2007 (Public Law 110-114), which authorized Indian River Lagoon-South, Picayune Strand Restoration, and the Site 1 Impoundment projects;
- Water Resources Reform and Development Act (WRRDA) 2014 (Public Law 113-121), which authorized four additional projects (C-43 Reservoir, C-111 Spreader Canal [Western], Biscayne Bay Coastal Wetlands [Phase 1], Broward County Water Preserve Areas [WPAs]); and
- WRDA 2016 (Title I of The Water Infrastructure Improvements for the Nation Act [WIIN Act]; Public Law 114-322), which includes authorization for the $1.9 billion Central Everglades Planning Project (CEPP). WRDA 2016 also authorized changes to the Picayune Strand Restoration Project related to cost escalations to allow for its completion.

A fourth WRDA was passed on October 10, 2018 (just prior to this report's release) and is expected to be signed into law by the president. WRDA 2018 authorized the 240,000 acre-foot (AF) Everglades Agricultural Area Storage Reservoir proposed in a CEPP post-authorization change report, which was developed by the SFWMD under Section 203 of WRDA 2000. No CERP projects are currently awaiting authorization. Thus, the authorization process does not pose any current delays on CERP restoration progress.

Authorized CERP projects are sometimes classified as Generation 1, 2, or 3 by the WRDA bills in which they were authorized, with the Melaleuca Eradication Project, which was authorized under programmatic authority, included in Generation 1. Section 1132 of WRDA 2016 included a requirement that the Secretary of the Army review completed post-authorization change reports and provide any recommendations to Congress within 120 days of completion of the report, which could potentially reduce administrative delays in CERP planning and authorization in the future.

Funding

Funding for Everglades restoration remains an important constraint on achieving a rate of progress that would be consistent with the original vision for the CERP. Federal funding for the CERP is illustrated in Figure 3-1, which include construction funds and support for planning, design, coordination, and monitoring. After a significant decrease to $44 million in fiscal year (FY) 2014, federal spending has increased to between $70-103 million over the 4 years FY2015-2018. The FY2019 USACE budget request includes $68 million for the CERP, amounting to 7.8 percent of the agency's construction budget (USACE, 2018a). Over the most recent 5-year period for which data are available (FY2014-2018), federal funding for Everglades restoration averaged $198 million per year, with

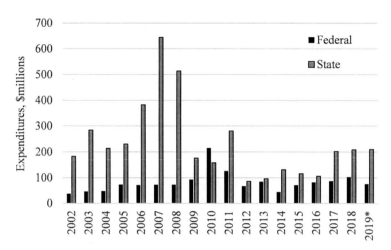

FIGURE 3-1 Federal and state funding for the CERP.

NOTE: Asterisk reflects budget requested.

SOURCE: SFERTF, 2018.

$77 million for CERP and $121 million for non-CERP efforts (Figure 3-2). These data show relatively steady CERP funding and variable non-CERP funding from the federal government.

State budgets for the CERP have sharply increased in recent years, while non-CERP funding has remained steady (Figures 3-1 and 3-2). From FY2014 to FY2018, state restoration spending averaged $153 million for CERP and $581 million for non-CERP efforts. In FY2017 and FY2018, CERP funding exceeded $200 million per year, nearly doubling spending levels over the previous 5 years.

South Florida Water Management District (SFWMD)'s Capital Improvements Plan for FY2018 through FY2022 reflects an intent to continue these higher levels of funding for the CERP. The Capital Improvements Plan calls for a 5-year total of $1.59 billion for surface water projects (an average of $318 million per year). About two-thirds of these projected expenditures are for CERP projects, and another 17 percent is for water quality improvements as part of the Restoration Strategies program (discussed later in this chapter).

The collection of previously authorized CERP projects combined with new projects still in various stages of planning (discussed in more detail later in this chapter) represent a very large financial commitment. Recent CERP project

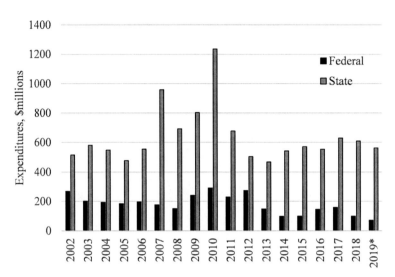

FIGURE 3-2 Federal and state funding for non-CERP restoration projects.

NOTE: Asterisk reflects budget requested.

SOURCE: SFERTF, 2018.

and program cost estimates, including only authorized projects, were approximately $12.3 billion (in 2016 dollars).[1] About $3 billion of that was funded through FY2016,[2] leaving a balance of roughly $9 billion. The Everglades Agricultural Area (EAA) Reservoir adds an increment of $1.3 billion, and the Lake Okeechobee Watershed Project is currently estimated to cost $1.4 billion. Two other projects in planning—the Western Everglades Restoration Project and the Loxahatchee River Watershed Project—do not yet have tentatively selected plans or cost estimates.

Estimation of annual funding requirements necessary to pay for these projects is dependent on the scheduling and final project costs. If federal funding for CERP projects continues at recent (5-year average) levels and is matched 50-50 by the state, about $155 million is available annually. At that rate, completion

---

[1] Reflects the $1.98 billion authorized cost of the Central Everglades Planning Project, authorized in 2016, and the $10.3 billion cost (in 2016 dollars) reported prior to authorization.

[2] NASEM (2016) reported $2.7 billion in state and federal CERP obligations as of FY2016. These obligations have not been inflation adjusted, so the number of $3 billion here is used as an approximation.

of currently authorized projects (including the EAA Storage Reservoir) would take approximately 65 years. With annual funding equal to FY2018 funding of $312 million annually, the timeline for completion of currently authorized projects is reduced to just over 30 years. Completion of projects in planning or currently unplanned CERP projects would lengthen that timeline. Increases in current funding levels would expedite those time horizons in a proportional manner.

Project Scheduling and Prioritization

The anticipated future progress of CERP projects and the relationships among all the federally funded South Florida ecosystem restoration projects and some highly relevant state-funded projects are depicted in the Integrated Delivery Schedule. The Integrated Delivery Schedule is not an action or decision document but rather a guide for planning, design, construction sequencing, and budgeting. The schedule is developed by the USACE and SFWMD in consultation with the South Florida Ecosystem Restoration Task Force and the many CERP constituencies. The Integrated Delivery Schedule replaced the Master Implementation Sequencing Plan, initially developed for the CERP, as required by the Programmatic Regulations (33 CFR §385.30).

An updated Integrated Delivery Schedule was released in July 2018 (USACE, 2018b). The reporting horizon for the 2018 schedule remains only through 2030 as in the previous December 2016 version. Modifications in the latest Integrated Delivery Schedule included numerous changes based on weather-related conditions, executions of contracts, and funding levels. The overall effect of these changes is a continued stretching out of the anticipated timeline for completing most of the authorized projects. Among the most significant programmatic changes is the acceleration of the Lake Okeechobee Regulation Schedule revision (previously to be completed in 2025, but now estimated in 2023) in response to the accelerated Herbert Hoover Dike rehabilitation schedule, with an estimated completion in 2022. The Integrated Delivery Schedule also shows that authorization is expected for the EAA Storage Reservoir in 2018; authorization of the Loxahatchee River Watershed Restoration and the Lake Okeechobee Restoration projects in WRDA 2020; and authorization of the Western Everglades Restoration Project in WRDA 2022. Once authorized, the new projects will create difficult political choices for the CERP program regarding whether to move forward with all projects simultaneously but more slowly or to prioritize some projects over others to expedite benefits.

As noted in NASEM (2016), the key limitations of the Integrated Delivery Schedule are that it is difficult to discern individual project costs or essential dependencies among projects; it does not include the full set of anticipated CERP projects (only project components scheduled through 2030 are included), giving a false impression on when the CERP will be completed; and there is no longer a requirement that RECOVER assess impacts of changes to the Integrated Delivery Schedule on expected CERP outcomes. The Programmatic Regulations required RECOVER to assess any changes in the master schedule "for effects on achieving the goals and purposes of the Plan and the interim goals and targets" (30 CFR 385.30[b][2]). Documentation of the effects of changes to the schedule on the nature and timing of anticipated ecosystem benefits would inform decisions to maximize regional benefits as soon as possible.

CERP RESTORATION PROGRESS

In the following sections the committee focuses on natural system restoration benefits emerging from the implementation of CERP projects (Figure 3-3). The project discussions are organized in order of authorization. For readers to understand the level of natural system restoration progress to be expected, a brief description of the state of implementation progress for each project is also provided in the text and in Table 3-1. The committee's previous report (NASEM, 2016) contains additional descriptions of the projects and progress up to October 2016, while this section emphasizes progress or new information gained during the past 2 years. The South Florida Environmental Report (SFWMD, 2018c), the 2015 CERP Report to Congress (USACE and DOI, 2016), and the 2016 Integrated Financial Plan (SFERTF, 2016) also provide detailed information about implementation and restoration progress. The following sections outline the natural system restoration progress based on monitoring to date at CERP projects for which construction has begun.

Generation 1 CERP Projects

Generation 1 projects are those authorized by Congress in WRDA 2007 (Picayune Strand Restoration, Site 1 Impoundment, and Indian River Lagoon–South) or by program authority (Melaleuca Eradication). A summary of implementation progress as of June 2018 is provided in Table 3-1. The location of the various projects is shown in Figure 3-3.

Picayune Strand Restoration

The Picayune Strand Restoration Project, the first CERP project under construction, focuses on an area in southwest Florida substantially disrupted by a real estate development project that drained 55,000 acres (about 86 mi^2) of

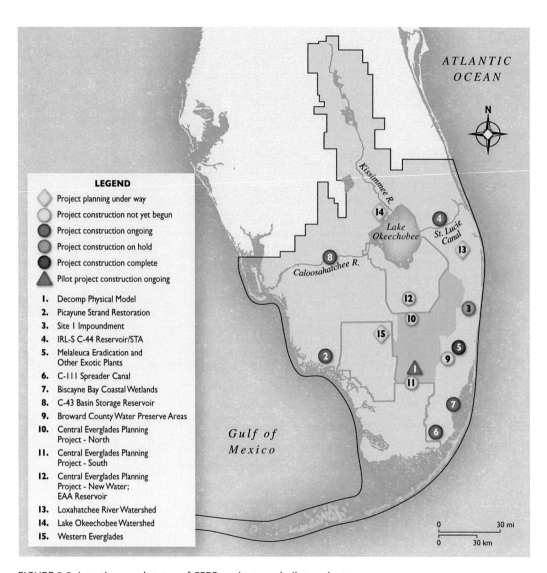

FIGURE 3-3 Locations and status of CERP projects and pilot projects.

SOURCE: © International Mapping Associates.

TABLE 3-1 CERP or CERP-Related Project Implementation Status as of July 2018

Project or Component Name	Yellow Book (1999) Estimated Completion	IDS 2018 Estimated Completion	Project Implementation Report Status	Authorization Status	Construction Status	Ecosystem Benefits Documented to Date
GENERATION 1 CERP PROJECTS						
Picayune Strand Restoration (Fig. 3-3, No. 2)	2005	2023	Submitted to Congress, 2005	Authorized in WRDA 2007	Ongoing	Increased water levels in 20,000 acres and with early vegetation responses detected
Site 1 Impoundment (Fig. 3-3, No. 3)						
- Phase 1	2007	2016	Submitted to Congress, 2006	Authorized in WRDA 2007	Completed, 2016	~16% reduction in seepage loss
- Phase 2		Not specified		Phase 2 requires further authorization.	Not begun	NA
Indian River Lagoon-South (Fig. 3-3, No. 4)			Submitted to Congress, 2004	Authorized in WRDA 2007		
- C-44 Reservoir/STA	2007	2020			Ongoing	None to date
- C-23/24 Reservoir/STA	2010	2030			Not begun	NA
- C-25 Reservoir/STA	2010	2027			Not begun	NA
- Natural Lands	NA	Not specified			Not begun	NA
Melaleuca Eradication and Other Exotic Plants (Fig. 3-3, No. 5)	2011	NA	Final June 2010	Programmatic authority WRDA 2000	Construction completed 2013, operations ongoing	Increased capacity for biocontrol

GENERATION 2 CERP PROJECTS						
C-111 Spreader Canal Western Project (Fig. 3-3, No. 6)	2008	2022	Submitted to Congress, 2012	Authorized in WRRDA 2014	Mostly complete; S-198 structure not yet constructed	Current data insufficient to assess response to project
Biscayne Bay Coastal Wetlands (Phase 1) (Fig. 3-3, No. 7)	2018	2022	Submitted to Congress, 2012	Authorized in WRRDA 2014	Ongoing	Some wetland vegetation responses to freshwater inputs; no change in nearshore salinity
C-43 Basin Storage: West Basin Storage Reservoir (Fig. 3-3, No. 8)	2012	2022	Submitted to Congress, 2011	Authorized in WRRDA 2014	Ongoing	None to date, construction ongoing
Broward County WPAs (Fig. 3-3, No. 9)			Submitted to Congress, 2012	Authorized in WRRDA 2014		
- C-9 Impoundment	2007	After 2030			Not begun	NA
- C-11 Impoundment	2008	2025			Not begun	NA
- WCA-3A & -3B Levee Seepage Management	2008	After 2030			Not begun	NA
GENERATION 3 CERP PROJECTS						
Central Everglades Planning Project (Fig. 3-3, Nos. 10, 11 and 12)	NA		Submitted to Congress, 2015	Authorized in WRDA 2016	Not begun	NA
- PPA South		2030				
- PPA North		2030				
- PPA New Water		After 2030				
EAA Storage Reservoir (Fig. 3-3, No. 12)	2015	NA	Submitted to Congress, 2018	Authorized in WRDA 2018	NA	NA

continued

TABLE 3-1 Continued

Project or Component Name	Yellow Book (1999) Estimated Completion	IDS 2018 Estimated Completion	Project Implementation Report Status	Authorization Status	Construction Status	Ecosystem Benefits Documented to Date
CERP PROJECTS IN PLANNING						
Loxahatchee River Watershed (Fig. 3-3, No. 13)	2013	NA	In preparation	NA	NA	NA
Lake Okeechobee Watershed (Fig. 3-3, No. 14)	2009-2020	NA	Draft PIR released July 2018	NA	NA	NA
Western Everglades (Fig. 3-3, No.15)	2008-2016	NA	In preparation	NA	NA	NA
REMAINING UNPLANNED CERP PROJECTS						
WCA Decompartmentalization (Phase 2)	2019	NA	NA	NA	NA	NA
Everglades National Park Seepage Management	2013	NA	NA	NA	Partly addressed by LPA Seepage Management Project	NA
Biscayne Bay Coastal Wetlands, Phase 2	2018	NA	NA	NA	NA	NA
C-111 Spreader Canal, Eastern Project	2008	NA	NA	NA	NA	NA
C-43 ASR	2012	NA	NA	NA	NA	NA
Site 1 Impoundment ASR	2014	NA	NA	NA	NA	NA
Agricultural Reserve Reservoir	2013	NA	NA	NA	NA	NA
North Lake Belt Storage Area	2021-2036	NA	NA	NA	NA	NA

Central Lake Belt Storage Area	2021-2036	NA	NA	NA	NA
WCA 2B Flows to Everglades National Park	2018	NA	NA	NA	NA
WPA Conveyance	2036	NA	NA	NA	NA
Caloosahatchee Backpumping with Stormwater Treatment	2015	NA	NA	NA	NA
West Miami-Dade Reuse	2020	NA	NA	NA	NA
South Miami-Dade Reuse	2020	NA	NA	NA	NA
A.R.M. Loxahatchee National Wildlife Refuge Internal Canal Structures	2003	NA	NA	NA	NA
Broward Co. Secondary Canal System	2009	NA	NA	NA	NA
Henderson Creek – Belle Meade Restoration	2005	NA	NA	NA	NA
Southern CREW	2005	NA	NA	NA	NA
Lake Trafford Restoration	2004	NA	NA	NA	NA
Southwest Florida Feasibility Studies	2004	NA	NA	NA	NA
Florida Bay Florida Keys Feasibility Study	2004	NA	NA	NA	NA
Comprehensive Integrated Water Quality Plan	2006	NA	NA	NA	NA

NOTES: Table 3-1 does not include non-CERP foundation projects. NA = not applicable. Remaining unplanned CERP projects include all projects more than $5 million (2014 dollars) as reported in USACE and DOI (2016), for which the components have not been incorporated in other planning efforts.
SOURCES: NASEM, 2016; USACE, 2018b.

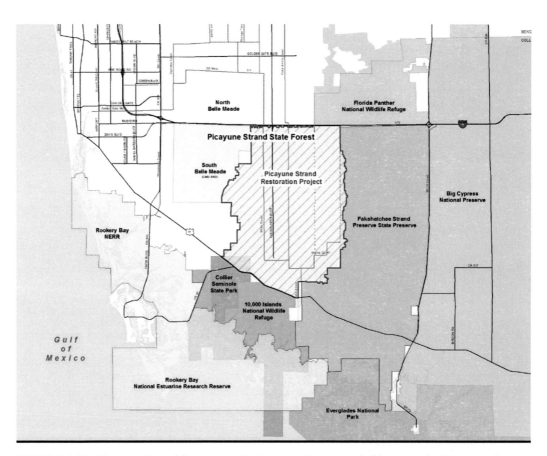

FIGURE 3-4 The Picayune Strand Restoration Project area is surrounded by several other natural areas, including Collier-Seminole State Park, Ten Thousand Islands National Wildlife Refuge, Picayune Strand State Forest, Fakahatchee Strand Preserve State Park, and Florida Panther National Wildlife Refuge. Restoration of water levels within the project footprint will enhance the hydrologic conditions in these surrounding natural areas.

SOURCE: Chuirazzi et al., 2018.

wetlands before being abandoned (Figure 3-3, No. 2). The roads and drainage disrupted sheet flow into Ten Thousand Islands National Wildlife Refuge, altered regional groundwater flows in surrounding natural areas, and drained a large expanse of wetland habitat (Figure 3-4).

The primary objective of the Picayune Strand Restoration Project is to "establish the pre-development hydrologic regime, including wet and dry season water levels, overland sheet flow, and hydroperiod" (RECOVER, 2014). An array of ecological objectives is dependent on this restoration of hydrology. Hydrologic restoration involves filling at least 50 percent of the length of the larger canals and several smaller ditches draining the area. The project also requires eliminating impediments to reestablishing sheet flow by removing raised roads and logging trams. There has been considerable progress in constructing the Picayune Strand Restoration Project, including canal plugging, road removal, and construction of pump stations (Table 3-2). The ecosystem responses expected to arise from hydrologic restoration include the reestablishment of natural plant distribution and composition, increase in fish and wildlife resources, restored habitat for listed species, and restored ecological connectivity to adjacent public lands (USACE and SFWMD, 2009). To achieve these benefits, the project requires not only the restoration of natural hydropatterns but also the control of exotic and nuisance plants and reestablishment of a natural fire regime in the Picayune Strand State Forest.

Because hydrologic restoration is a pre-condition for ecological restoration, hydrologic monitoring should provide the first signals of potential restoration success. A robust monitoring effort for both hydrologic and ecological objectives (USACE and SFWMD, 2009; see also Chapter 4) has been established. The hydrologic monitoring results are then utilized to delineate the project area into three levels of hydrologic restoration achieved to date—full, partial, and no hydrologic restoration—determined based on the project components constructed and the local influences of neighboring canals on water levels (see Figure 3-5). Since construction has begun, ecological monitoring has been focused on the areas with full or partial hydrologic restoration, utilizing reference sites in neighboring Fakahatchee Strand Preserve State Park and Florida Panther National Wildlife Refuge (Barry et al., 2017; Worley et al., 2017).

Patterns of lengthening hydroperiod (the period of time per year that water levels are at or above ground surface) appear to follow restoration progress, with the largest gains demonstrated at those wells in the fully restored areas. For example, Figure 3-6 shows evidence of lengthening hydroperiod in the upper reach of the Prairie Canal (Well 10 in Figure 3-5) almost immediately after plugging the Prairie Canal (2004 through 2007) and further lengthening after the filling of the neighboring Merritt Canal in 2015, when the area was considered to be fully restored. Records from this area for 1997-2004 indicated a lack of recorded standing water, in contrast to the significant hydroperiods observed in 2008, 2013, fall and winter of 2015, and 2016. In contrast, Figure 3-7 depicts a partially restored area along the Merritt Canal (Well 8 in Figure 3-5), which appears to be responding to the canal plugging and removal of roads and logging trams completed in 2015. This area had no significant periods of inundation

TABLE 3-2 Phases and Progress of the Picayune Strand Project

	Lead Agency	Road Removal (mi)	Logging Tram Removal	Canals to Be Plugged (mi)	Other	Project Phase Status
Tamiami Trail Culverts	State	NA		NA	17 culverts constructed	Completed in 2007
Prairie Canal Phase	State (expedited)	64	30	7	Hydrologic restoration of 11,000 acres in Picayune Strand and 9,000 acres in Fakahatchee Strand State Preserve Park	Plugging and road removal completed in 2007; logging trams removed in 2012
Merritt Canal Phase	Federal	65	16	8.5	Merritt pump station, spreader basin, and tie-back levee constructed	Completed in 2015; pump station transferred to SFWMD in 2016
Faka Union Canal Phase	Federal	81	11	7.6	Faka Union pump station, spreader basin, and tie-back levee constructed	Roads removed in 2013; pump station completed in 2017; canal plugging scheduled for 2022
Miller Canal Phase	Federal/State	77	11	13	Construct Miller Canal pump station, spreader basin, tie-back levee, and private lands drainage canal; remove western stair-step canals	Miller pump station completed May 2018; road removal and canal plugging scheduled for 2019 and 2022, respectively
Manatee Mitigation Feature	State	0	0	0	Construct warm water refugium to mitigate loss of existing refugium	Completed in 2016
Southwestern Protection Feature	State	0	0	0	Construct 7-mile levee for flood protection of adjacent lands	Construction completion scheduled for 2022
Stair-step canals between Prairie and Faka Union Canals	Federal	0	0	5.2		Construction completion estimated in 2018

SOURCE: J. Starnes, SFWMD, personal communication, 2016.

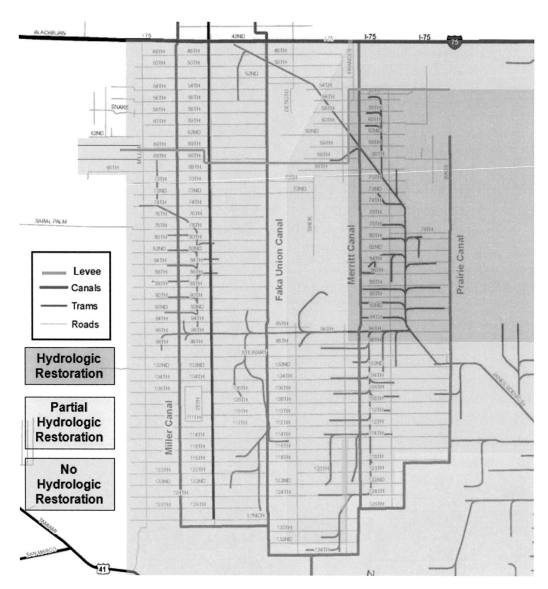

FIGURE 3-5 Schematic approximation of hydrologic restoration at Picayune Strand with locations of vegetation monitoring transects and monitoring wells for the 2016 sampling event.

SOURCE: Chuirazzi et al., 2018.

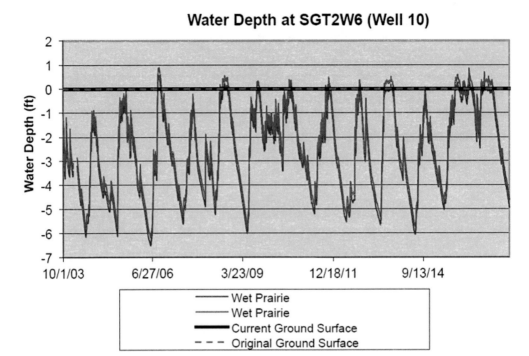

FIGURE 3-6 Groundwater depth at the Prairie Canal, which achieved partial hydrologic restoration in 2004 and 2007 and full hydrologic restoration in 2015. The red and blue lines represent two wet prairie vegetation sampling sites near Well 10.

SOURCE: Barry et al., 2017.

until 2015 and 2016, in which the completion of restoration activities coincided with above average rainfall years. Although the hydrographs visually indicate that restoration is progressing, the analysis of the information is only qualitative and lacks rigorous statistical comparisons of the restored areas against the reference sites (see Chapter 4).

Vegetation transects have been sampled six times since 2005 (see Chapter 4 for more details). The most recent 2016 vegetation monitoring (Barry et al., 2017) describes the results for vegetation transects in three types of pre-drainage habitat—cypress, wet prairie, and pineland—which are reported by four strata (i.e., vertical layers of vegetation): canopy trees, subcanopy trees, the

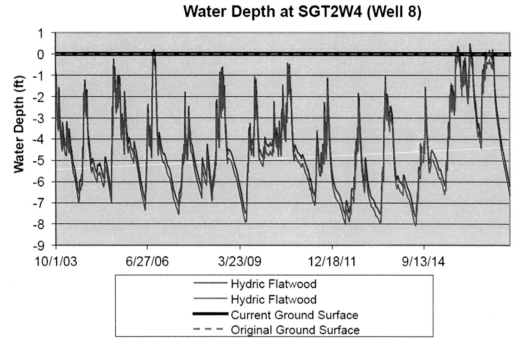

FIGURE 3-7 Groundwater depth at the Merritt Canal, which achieved partial hydrologic restoration in 2015. The red and blue lines represent two hydric flatwood vegetation sampling sites near Well 8.

SOURCE: Barry et al., 2017.

shrub layer, and groundcover.[3] It should be expected that time to response will differ by habitat and strata; for example, standing trees in cypress habitat may not have a visible response for decades, while groundcover in freshwater marsh habitat may respond within a growing season or two. Thus, the 2016 monitoring may not have allowed an appropriate lag time for evidence of a response in all habitat types.

The vegetation monitoring results provide an emerging trajectory of restoration, although few definitive conclusions can be drawn from the data. Across all habitat types, short-term changes are most evident in the groundcover stratum, as

---

[3] Canopy trees consist of woody plants with a diameter at breast height (dbh) greater than 10 cm; subcanopy trees consist of woody plants with a dbh between 2.5 and 10 cm, excluding woody shrubs; the shrub layer consists of trees with a dbh of less than 2.5 cm and all shrub individuals; and groundcover consists of all remaining plants and primarily herbaceous species.

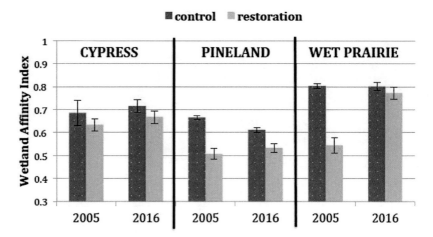

FIGURE 3-8 Groundcover wetland affinity index comparison of 2005 and 2016. Marsh and hammock habitat types were not included because of data limitations.

NOTE: "Control" here describes the reference sites.

SOURCE: Barry et al., 2017.

should be expected. Comparisons of the groundcover in the wet prairie habitat between 2005 and 2016 show significant recovery. By 2016, wet prairie Wetland Affinity Index values (a measure of the probability that the observed species generally occur within wetlands) were similar to reference transects, which represent the target restoration conditions (Figure 3-8). In contrast, restored transects in pinelands have not shown appreciable improvement, although this is partially explained because they occur in areas with partially restored hydrology.

Faunal indicators (e.g., treefrogs, aquatic macroinvertebrates, and fish) were sampled for the first time post-restoration in 2016. The monitoring data show a system that is generally in transition from short hydroperiods to more sustained surface water inundation, but with a high variability between individual sites (Worley et al., 2017), which makes significant differences between restored, partially restored, and nonrestored sites more difficult to detect (see Chapter 4). Macroinvertebrates sampling showed no significant difference in diversity of species between restoration and reference sites. A slight indication of a macroinvertebrate shift at restored sites can be seen between the 2005-2006 and 2016-2017 samplings and may be the beginning signal of a temporal shift.

The aquatic fauna monitoring data provide striking evidence of the challenge of invasive species. For example, the total treefrog population across all sites consists of only three species, with an exotic species (the Cuban treefrog) accounting for 98 percent of the individuals captured. There was no significant difference in treefrog composition between reference and restored sites, and the data indicate that Cuban treefrogs are outcompeting native species across all sites. Fish composition showed a similar challenge, with the post-restoration study documenting the presence of the non-native African jewelfish across both reference and restored sites.

In general, because of the ongoing construction and extended construction timeline with estimated completion in 2023, evaluations of restoration success at Picayune Strand require assessment by the level of hydrologic restoration. In areas considered fully restored, early signals of hydrologic restoration are provided by well data showing increasing hydroperiods, and vegetation response is evident in groundcover strata with increasing Wetland Affinity Index values that show similarity between reference targets and restored sites. However, few statistically significant responses to restoration have been documented. In Chapter 4, the committee discusses improved monitoring and assessment strategies to more rigorously demonstrate early restoration success.

Site 1 Impoundment

The Site 1 Impoundment Project (No. 3 on Figure 3-3; Figure 3-9) was originally cast as a single-phase effort to modify local hydrologic conditions to store more water (13,300 AF) and to help alleviate demands on water in Arthur R. Marshall Loxahatchee National Wildlife Refuge (LNWR). Without the project, during wet periods, runoff from LNWR is shunted to the ocean, while during dry periods, water is taken from the LNWR to meet user demands elsewhere. Planners projected that the Site 1 impoundment would allow for better management of water to supply natural system demands within the LNWR (USACE and SFWMD, 2016a). In 2009, the project was divided into two phases (see Figure 3-9). Construction of Phase 1, completed in 2016, was an $81 million effort that included modifications to the existing L-40 levee and construction of a 6-acre wildlife wetland area (USACE and SFWMD, 2016a; G. Landers, USACE, personal communication, 2016). Phase 1 is estimated to provide a 16 percent reduction in existing seepage at the L-40 levee (USACE, 2016a). Phase 2 of the project requires further congressional authorization necessitated by increased costs (USACE, 2018c). The SFWMD, however, in 2016 communicated to the USACE that it is no longer interested in constructing Phase 2, because of the high anticipated cost of the plan relative to the benefits provided (M. Morrison, SFWMD, personal communication, 2016). CERP planners have not formally deauthorized the project. However, based on current agency support and the fact

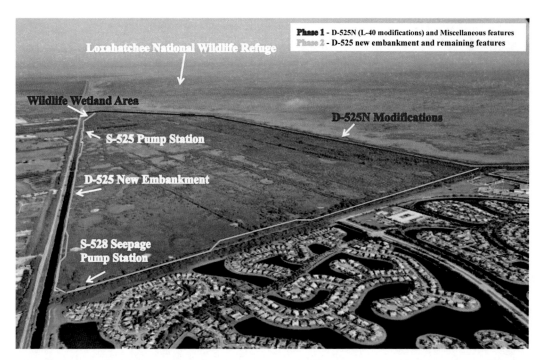

FIGURE 3-9 Location of the Site 1 Impoundment project, looking west-northwest. The Arthur R. Marshall Loxahatchee National Wildlife Refuge is at the upper right of the image. As with many restoration projects, Site 1 has a sharp boundary between its restoration area and neighboring urban development.

SOURCE: Modified from Audubon of Florida, 2010.

that the project was deleted from the July 2018 draft Integrated Delivery Schedule (USACE, 2018b), it appears unlikely that Phase 2 of the Site 1 Impoundment Project will ultimately be constructed.

Indian River Lagoon-South

The Indian River Lagoon and St. Lucie Estuary are biologically diverse estuaries located on the east side of the Florida Peninsula, where ecosystems have been impacted by polluted runoff from farmlands and urban areas and surges of freshwater (USACE, 2013). The Indian River Lagoon-South (IRL-S) Project (Figure 3-3, No. 4) is designed to reverse this damage through improved water management, including the 50,600 AF C-44 storage reservoir, three additional reservoirs with a total of 97,000 AF of storage, four new stormwater treatment areas (STAs), dredging of the St. Lucie River to remove 7.9 million cubic yards of muck, and restoring 53,000 acres of wetlands, among other features. The project is anticipated to cost $3 billion in 2014 dollars (USACE and DOI, 2016). Construction is under way on the C-44 reservoir and STA, with estimated completion in 2020 (Gonzales, 2018).

Melaleuca Eradication and Other Exotic Plants

The Melaleuca Eradication and Other Exotic Plants Project is a CERP effort to address the potential threat to restoration posed by non-native invasive plant species. Five invasive species that are particularly problematic are the focus of major ongoing management efforts: Melaleuca (*Melaleuca quinquenervia*), Brazilian pepper (*Schinus terebinthifolius*), Australian pine (*Casuarina* spp.), old world climbing fern (*Lygodium microphyllum*), and air potato (*Dioscorea bulbifera*). A crucial part of this work is centered at the U.S. Department of Agriculture's Invasive Plant Research Laboratory in Davie, Florida, where specific biological control agents—mostly insects—are developed. With CERP funds, the U.S. Department of Agriculture has constructed a 2,700-square-foot annex to the present laboratory to facilitate additional mass rearing (Figure 3-3, No. 5). The annex was completed in 2013 and has been transferred to the local sponsor (USACE, 2015a). The project includes CERP operations and maintenance funding for mass rearing, release, and field monitoring of biocontrol agents to manage the spread of invasive non-native plant species in the Everglades and South Florida (USACE and SFWMD, 2015a).

The Invasive Plant Research Laboratory releases biocontrol agents for invasive plants widely in South Florida, including in Everglades National Park (IPRL, 2017). The number of releases has been increasing, with more biocontrol agents released in 2017 than in any previous year. A total of 283,000 air potato beetles (*Lilioceris cheni*), 1,600,000 waterhyacinth planthoppers (*Megamelus scutellaris*), 1,480,000 Lygodium moths (*Neomusotima conspurcatalis*), and 742,000 Lygodium mites (*Floracarus perrepae*) were released in the four years 2014-2017. Releases of the air potato beetle have reduced the growth rate and spread of air potato in many areas in the Everglades ecosystem. Research has shown that the Lygodium mite can complement fire control efforts by severely impacting regrowth, while the waterhyacinth planthopper has been shown to make the plants more susceptible to herbicides, possibly offering a path to reduced herbicide application without loss of control efficacy (IPRL, 2017).

The Melaleuca Eradication and Other Exotic Plants Project is one effort among many efforts to control invasive plant species in the Everglades, and several federal and state agencies are engaged to control these problem plants.

As a result, the exact contributions of this CERP project in the overall effort are difficult to parse, although the control of invasive plants is essential to achieve restoration goals (NRC, 2014).

Generation 2 CERP Projects

Four second-generation CERP projects were authorized as part of WRRDA 2014 (Table 3-1): the Biscayne Bay Coastal Wetlands (Phase 1) Project, the C-111 Spreader Canal (Western) Project, the C-43 Reservoir, and the Broward County Water Preserve Areas. No construction has begun on the Broward County Water Preserve Areas, so the discussions will focus on the other three projects.

Biscayne Bay Coastal Wetlands (Phase 1) Project

The Biscayne Bay Coastal Wetlands Project (Figure 3-3, No. 7) is designed to address near-shore hypersalinity and to improve the ecological condition of the wetlands, tidal creeks, and near-shore habitat by restoring the timing and quantity of freshwater flows into the bay and Biscayne National Park. Historically, drainage and development cut off the wetlands from their source of freshwater, resulting in wetland losses and an increase in salinity along the margin of the bay. The overall project seeks to reverse these effects on 11,300 acres of the total 22,500 acres of wetlands. Phase 1 of the Biscayne Bay Coastal Wetlands Project is small, with a total footprint of approximately 3,800 acres divided into three geographically distinct regions: the Deering Estate Flow-way, the Cutler Wetlands Flow-way, and the L-31E Flow-way (SFERTF, 2016; Figure 3-10). The project includes the construction of pump stations, spreader canals, and culverts and the restoration of flow-ways (USACE, 2018d). Construction of the Deering Estate component and 4 of the 10 culverts in the L-31E Flow-way were expedited by the SFWMD and completed by 2012. By 2018, the last 6 culverts were completed, and in April 2017, the SFWMD installed an interim pump to increase water available to flow through the culverts. The USACE is expected to construct the remaining features of the L-31E Flow-way, including four pump stations, by 2022. Construction of the Cutler Wetlands component is scheduled for 2020-2021 (USACE, 2018d).

Deering Estate. The S-700 pump station on the C-100A Spur Canal within the Deering Estate is designed to restore historic freshwater flow through the wetlands and into Biscayne Bay to reduce near-coastal salinity. In water year (WY) 2017, the pump station diverted approximately 12,900 AF of freshwater from the C-100 Canal to the remnant wetlands near Cutler Creek (Charkhian et al., 2018).

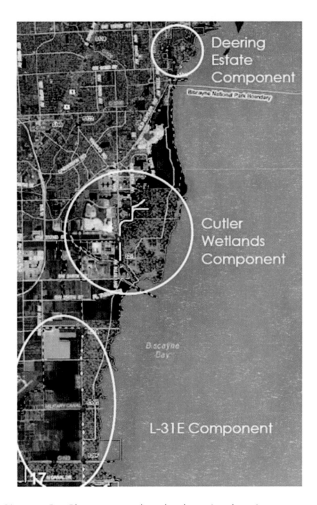

FIGURE 3-10 Biscayne Bay Phase 1 coastal wetlands project locations.

SOURCE: USACE and SFWMD, 2012.

Although this component has a small footprint, the project is seeing some progress toward the project goals. Since pumping began, salinity in Cutler Creek and in groundwater has decreased. Salinity in the upper reaches of Cutler Creek, which was subject to saltwater intrusion, decreased to less than 1 practical salinity unit (psu)[4] in response to freshwater inputs. Groundwater stage has

---

[4] Psu is nearly equivalent to parts per thousand but is measured using electrical conductivity.

risen, and groundwater salinity, which typically peaked at more than 20 psu in the dry season under the coastal mangroves, was reduced to less than 10 ppt by freshwater inputs (Charkhian et al., 2017, 2018). Nearshore salinity has not shown as much response to the project and remained on average above the RECOVER target of 20 psu.

In the initial design, the Deering Estate S-700 pump station would be controlled by a programmable logic controller based on the stage, but this functionality was prevented by programming limitations. Instead, the District's operations control room has controlled pumping, and the project has largely been operated under discontinuous pumping. The permeable limestone bedrock allows the water level to drain quickly when pumping stops, leading to wide fluctuations in water levels and periods with no standing water. Recent pump tests to investigate the effects of a pulsed versus continuous pumping schedule resulted in a RECOVER recommendation for a constant pumping rate of at least 25 cubic feet per second (cfs) (representing at least 19 acres inundated, or 58 percent of the historic wetlands in the project area). Constant pumping reduced atypical water level fluctuations and improved wetland salinity, although the area of inundation was reduced (Charkhian, 2017).[5] The revised operations will be implemented in WY2019 (M. Jacoby, SFWMD, personal communication, 2018).

The inability to date to create a natural hydroperiod may have limited the ecological response. For example, the project has not yet facilitated the expected shift in the vegetation communities, although there is some evidence of a shift toward hardwood tree species that are more tolerant of flooding (Charkhian et al., 2018). Some upland vegetation that had encroached into the wetland has begun to die back with mean vegetation canopy cover declining slightly between 2013 and 2016, although the differences were not statistically significant (Charkhian, 2017).

Although these operations represent improvements, it is not clear that the low pumping rate will provide the ecological benefits that are predicted for this component of the Biscayne Bay Coastal Wetland Project. A reassessment of the predicted ecological outcomes under the new pumping rates is warranted (e.g., near-shore salinity, vegetation response) in order to determine whether the expected ecological targets can be met. This presents an opportunity to learn more about the system's response and to implement adaptive management options as appropriate (USACE and SFWMD, 2011a).

The L-31E Component. The L-31E Culverts component aims to improve habitat conditions by diverting water from canal discharges into coastal wetlands, thereby

---

[5] At a rate of 100 cfs, 94 percent of the historic wetlands or 31 acres are inundated (Charkhian, 2017).

improving near-shore salinities. Delivering sufficient freshwater through the L-31E Culverts to wetlands east of the L-31E Canal has been an ongoing challenge because of the lack of pumps to move water to the culverts and hold the stage high enough in the canal to promote flow into the coastal wetlands. The SFWMD reports a regulatory-based performance target to divert 4 percent of the total coastal discharges to the wetland as freshwater flow from the L-31E Canal,[6] although none of the project documents demonstrates that this target flow will meet the project's ecological objectives. RECOVER (2014) states that "the volume of water diverted may provide a more direct restoration-based target than the percentage of available water." Between water years[7] 2012 and 2017, the project has only met the regulatory 4 percent flow target in 24 of the 84 months (Figure 3-11). Dry season flows rarely occurred outside of periods when the temporary or interim pumps were in place, and even in wet seasons, the stage in the L-31E Canal is lower than the "optimal level" necessary to drive the target freshwater flows into the coastal wetland (Charkhian, 2017). Dry season flows are anticipated to increase when the four pump stations are constructed (anticipated in 2020).

Sawgrass mapping within a 370-acre area of the Miami Dade-County preserve wetlands in February 2017 showed a marginal increase in its areal extent, increasing from 43.1 to 52.0 acres between 2013 and 2017 and replacing upland species. There has also been a slight increase in the abundance of various bird species, amphibians, invertebrates, and fish species relative to baseline abundance data (Charkhian et al., 2018), although these reports are anecdotal and not linked to a monitoring performance target. Given that the hydrologic performance targets have not been met, it is not surprising that the ecological response has been limited.

Across both project components, there is a surprising lack of rigorous analysis of the monitoring data relative to project objectives and expectations, given the time that the project components have been operational. Without this analysis, it is difficult to clearly assess restoration progress relative to expectations and whether adaptive management steps are needed.

C-111 Spreader Canal (Western) Project

The C-111 Canal (Figure 3-3, No. 6) is the southernmost canal for the entire Central and Southern Florida Project. Originally designed to provide flood protection for agricultural lands to the east in Dade County, a major problem resulted when the C-111 Canal also drained water from the Southern Glades and Taylor Slough in Everglades National Park. Much of the water in the canal is a result of

---

[6] Special Conditions 10-C of USACE Permit SAJ-2007-1994-1327 [IP-TKW].
[7] The water year runs from May 1 through April 30.

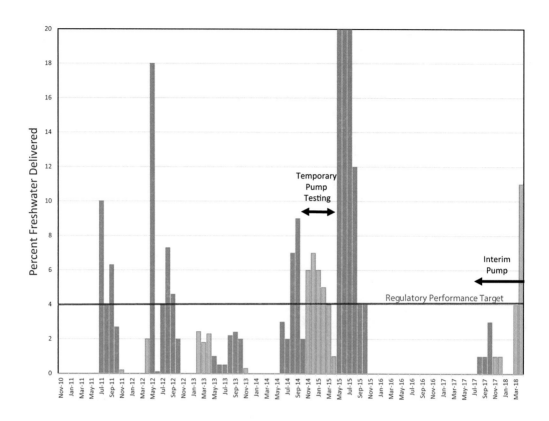

FIGURE 3-11 Percentage of total freshwater deliveries diverted to coastal wetlands via L-31E Culverts relative to 4 percent regulatory-based performance target. Orange bars are deliveries in the dry season; blue bars are deliveries in the wet season.

SOURCE: Data from Charkhian et al., 2014, 2015, 2016, 2017a, 2018; M. Jacoby, SFWMD, personal communication, 2018.

seepage from Everglades National Park to the west. The amount of seepage and the resulting change in flow pattern caused environmental and ecological damage to Taylor Slough, which became too dry; at the same time, Barnes Sound and Manatee Bay suffered ecological damages as high freshwater flows upset the natural salinity balance of their waters. Working in concert with the non-CERP C-111 South Dade Project and the SFWMD Florida Bay Initiative to the north (discussed later in this chapter), the C-111 Spreader Canal (Western) Project promises to restore the volume, distribution, and timing of flow into Taylor Slough and to improve salinity regimes in eastern Florida Bay (SFWMD, 2013).

The C-111 Spreader Canal Project is structured in two phases. The first phase—the Western Project—creates a 6-mile-long hydraulic ridge along the eastern boundary of Everglades National Park to reduce seepage from the park and to improve the hydrology of Taylor Slough and includes canal modification to reduce canal flows into eastern Florida Bay. The Western Project features include two pump stations (S-199 and S-200), a 516-acre detention basin (the Frog Pond Detention Area), canal plugs, and the Aerojet Canal impoundment and weirs (see Figure 3-12). The C-111 Spreader Canal (Western) Project was largely completed in February 2012 and began operations in June 2012. One additional new structure (S-198) is authorized in the lower section of C-111, which has not yet been scheduled for construction (USACE, 2018e). The second phase of the C-111 Spreader Canal Project (Eastern phase) has not yet been specifically planned or authorized.

The objective of the CERP project is not to add water to Taylor Slough but rather to create a hydraulic ridge that prevents water flowing in Taylor Slough

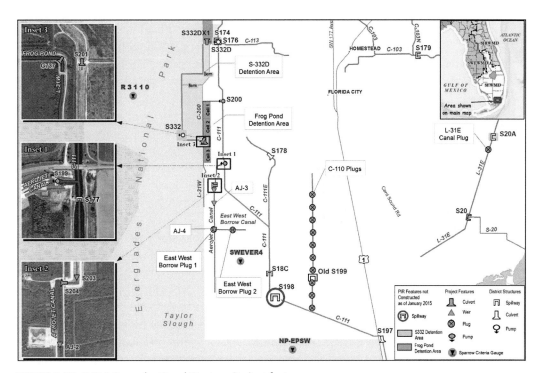

FIGURE 3-12 C-111 Spreader Canal Western Project features.

SOURCE: Qui et al., 2018.

from seeping into the C-111 canal. Water pumped into the Frog Pond Detention Area (through S-200; see Figure 3-12) and the Aerojet Canal impoundment (through S-199) will seep into the ground and back into the canal. The expectation is that the reduction in seepage from Taylor Slough will increase flow into Florida Bay, reducing salinity and having positive ecological effects. The hydrologic ridge element of the project only functions when water is available to fill the detention areas.

The natural system benefits of the C-111 Spreader Canal (Western) Project are difficult to completely separate from those of neighboring non-CERP projects, including the C-111 South Dade Project and the Florida Bay Initiative. The S-332D Detention Area of the C-111 South Dade Project (see Figure 3-12) has been operating since 2003 and has similar overarching objectives (D. Crawford, USACE, personal communication, 2018). The Florida Bay Initiative has been operating since 2017. The results shown here are assumed to be the collective response of multiple projects. Two reports that evaluate the monitoring data associated with project success are Qui et al. (2018) and Kline et al. (2017).

Qui et al. (2018) focuses on water year 2017 and indicates that the project has been operating generally as expected. Annual mean flow was 82 cfs at S-199 and 113 cfs at S-200, with operations primarily in the wet season. The report summarizes data on hydroperiod in the Southern Glades and Model Lands and salinity in the coastal zone but does not draw conclusions about project effectiveness.

Kline et al. (2017) provide a more detailed analysis of data for water year 2016 based on an analysis of data from 1993 to 2016. They estimate the amount of water that flows through Taylor Slough, into the Southern Glades between S-198 and S-197, and directly into eastern Florida Bay (Manatee Bay) via canal discharge (through S-197). The authors also estimate seepage back into the canal to the east of the Aeroject Canal feature (calculated as the difference in flow from S-177 to S-18C).

These hydrological responses over time are partially described in Figure 3-13, which shows substantial year-to-year variability. In years prior to project operation, a significant amount of water flowed to Manatee Bay (part of the Biscayne Bay watershed) via C-111 Canal discharge and over land from the C-111 Canal into the Southern Glades to northeastern Florida Bay, causing these areas to have lower salinities, while less water flowed through Taylor Slough to north central Florida Bay (see Figure 3-13). Since project operations started in 2012, some years showed increased Taylor Slough flows relative to eastern discharges (2013, 2014) and other years seemed to show little effect (2012, 2015). Based on a detailed seasonal analysis of the water budget for C-111SC and rainfall, Kline et al. (2017) conclude that the project can successfully increase Taylor Slough

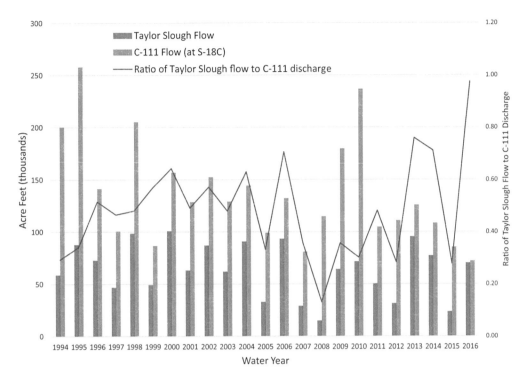

FIGURE 3-13 Comparison of annual freshwater flow volumes into Taylor Slough (orange bars) from 1993 and 2016 measured at the Taylor Slough Bridge compared to C-111 discharges to Florida Bay measured at S-18C (gray).

SOURCE: Modified from Kline et al., 2017.

flows to Florida Bay if rainfall is sufficient, but the project is less effective during periods of drought because water for project operations is insufficient. Kline et al. (2017) conclude that additional water will be needed to meet the project goals. Additional water may be available to the project through the implementation of the CEPP and the Combined Operations Plan for Mod Waters and C-111 South Dade, but these operational decisions have not been determined.

Kline et al. (2017) also estimated the efficiency of the project by calculating the amount of seepage back into the canal after water enters Taylor Slough through the S-200 and S-199 pumps (the efficiency is the estimated percentage not lost to seepage). They estimated that the S-199 and S-200 pumps combined were only 7 percent efficient in the wet season compared to 73 percent in the dry season for water year 2016. The higher dry season efficiency is likely due to a steeper westward hydraulic gradient when marsh water levels are low. The benefits provided from the project could be even greater in the wet season if seepage could be further reduced.

Most of the ecological monitoring associated with C-111 Spreader Canal has focused on responses in Florida Bay. However, there is some evidence of increases in long-hydroperiod species such as *Eleocharis* (spikerush) in Taylor Slough (Figure 3-14). Troxler et al. (2013) suggest that with sustained water management, shifts in plant community structure could occur in 5-10 years. The effects of the project on ecological metrics in Florida Bay and near-shore habitats are difficult to assess for two reasons. First, the expected effect of the project on salinity is small relative to natural variation—Qui (2016) notes that project modeling indicates a net improvement in salinity concentrations of 3 percent in near-shore embayments. Second, sea-level rise has resulted in saltwater intrusion, potentially masking the effects of the project on submerged aquatic

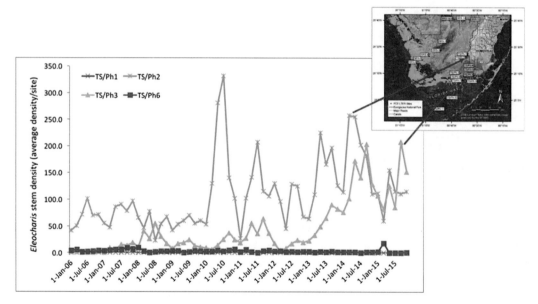

FIGURE 3-14 High-quality marsh habitat macrophyte species *Eleocharis* expanding at two sites downstream of C-111 Spreader Canal operations since 2012.

NOTE: The Ph1 and Ph6 sites did not have *Eleocharis*. The Ph1 site has never had *Eleocharis*, and the Ph6 site has salinity levels that adversely affect *Eleocharis* growth.

SOURCE: F. Sklar, SFWMD, personal communication, 2018.

vegetation (SAV) and freshwater fish abundance. The past four hydrologic years have seen high water levels during the dry season, which resulted in a low number of days where the water level associated with high prey concentrations was optimal. The changes in water levels associated with sea-level rise also affected roseate spoonbill nesting success as feeding is connected with low water levels that concentrate freshwater fish prey. Kline et al. (2017) conclude that "sea level rise is having profound effects on the upper trophic levels in the Florida Bay coastal habitats that may alter how restoration projects are assessed."

Based on available monitoring data, the hydrologic and ecological effects of the project have not been fully determined. Extreme events in 2015-2017, including severe drought and seagrass die-off, have made an evaluation of progress difficult. These events have led to encroachment of marine water into transitional creeks, movement of freshwater fish upstream, changes in avian ecology, and possible increases in nutrients (Kline et al., 2017; Sklar and Dreschel, 2018), which may confound the assessment of project effects. Monitoring to evaluate project success is discussed in more detail in Chapter 4.

C-43 Storage Reservoir

A major environmental issue in the estuary of the Caloosahatchee River on the west coast of Florida is the restoration and maintenance of appropriate salinity levels for aquatic organisms, particularly shellfish. Early in the 20th century, the course of the Caloosahatchee River was deepened and straightened, and canals were dug in the river basin to provide a capacity for drainage of agricultural lands and urban areas. As a result, during prolonged dry periods, freshwater flow to the estuary is greatly reduced, to the extent that saline water can migrate far up the river and kill beds of freshwater submerged plants. During periods of heavy rainfall, large volumes of nutrient- and sediment-rich freshwater are transported into the estuary, affecting habitat quality for seagrasses, oysters, and other aquatic organisms. The Caloosahatchee River (C-43) West Basin Storage Reservoir (Figure 3-3, No. 8) is a CERP project designed to impound up to 170,000 AF of stormwater runoff from the C-43 drainage basin or from Lake Okeechobee during wet periods (USACE and SFWMD, 2016b), hence protecting the estuary from excessive freshwater. During dry periods, this stored water can be released to supplement low river flows to maintain optimal salinity levels in the estuary and is available for water supply. The project has four phases associated with construction. The first phase of construction began in late 2015, and the reservoir is anticipated to be completed by July 2022 (SFWMD, 2017). Because construction is still in an early phase, it is too soon to see natural system benefits from this project.

Generation 3

A single large project—the CEPP (Figure 3-3, Nos. 10, 11, and 12; NRC, 2014; USACE and SFWMD, 2014)—makes up Generation 3. CEPP was authorized in WRDA 2016 as a $1.9 billion project. CEPP includes a flow equalization basin to provide storage to increase the quantity of water flowing into the remnant Everglades and improve water quality treatment for the new water. The project also improves the distribution of flow through seepage management, the filling of canals, and the levee removal in the central Everglades. The project has been divided into three phases based on project partnership agreements (PPAs): PPA North, PPA South, and PPA New Water. Construction has not begun, and therefore there are no project-related benefits to discuss, although construction of two early components of PPA South (the S-333N pump station and removal of old Tamiami Trail) are expected to begin in October 2018.

A post-authorization change report (SFWMD, 2018a) submitted by the SFWMD to convert the A-2 flow equalization basin to a 240,000 AF reservoir was authorized by Congress in October 2018. The committee examined this project component but did not rigorously review the project analysis or the selection of alternatives.

EAA Storage Reservoir Project

In May 2017, the Florida legislature enacted Water Resource Law of 2017 (Senate Bill 10), which mandated a new planning effort focused on water storage in the Everglades Agricultural Area. This project was intended to reduce harmful freshwater discharges from Lake Okeechobee to the St. Lucie and Caloosahatchee estuaries and to increase the flow of freshwater to the Everglades. The Florida Senate appropriated $800 million for the planning, design, and construction of the project. The law specifically directed the SFWMD to develop a post-authorization change report for the CEPP that evaluates the benefits of a potential reservoir with a minimum of 240,000 AF storage along with required associated treatment on the footprint of the proposed 56,000 AF A-2 flow equalization basin (FEB) as authorized in CEPP. The law allowed for consideration of an alternate configuration using the Restoration Strategies A-1 FEB if a minimum of 360,000 AF total storage could be provided with the necessary treatment. The legislature authorized the SFWMD to acquire additional land for the project from willing sellers but specifically prohibited the use of eminent domain. The law set a stringent timeline for the planning process, requiring that an approved report be submitted to Congress for authorization by October 1, 2018. The SFWMD conducted numerous public meetings during the planning process to ensure consistency with the National Environmental Policy Act and eligibility for federal cost sharing and delivered the post-authorization change report to the USACE on March 26, 2018. The project was conditionally approved by the Office of Management and Budget in July 2018.

In its development of the tentatively selected plan, the SFWMD evaluated five initial alternatives, representing two basic configurations and footprints, with different locations of the reservoirs and STAs within the footprints. One alternative (C360C) used a different operational plan to allow for water supply. The SFWMD used the Regional Simulation Model under contemporary (1965-2005) climate conditions to analyze future benefits and assumed that Lake Okeechobee would operate under its current regulation schedule with the additional flexibility assumed in the CEPP. In addition, the analysis excluded the potential benefits of other projects in planning as well as the potential impacts of sea-level rise and changes in precipitation and temperature patterns. After further evaluation and optimization, the SFWMD identified a final alternative (C240A) that enables water supply deliveries from the reservoir. The proposed plan is comprised of a 240,000 AF, 23-foot-deep reservoir with multipurpose operational flexibility, a 6,500-acre STA, and conveyance improvements. The new reservoir would take the place of the Central Everglades A-2 FEB with additional land from willing sellers to the west, preserving the continued functioning of the Restoration Strategies A-1 FEB. In addition, the plan includes conveyance improvements to the Miami and North New River Canals. The net increase in the cost of the CEPP based on the post-authorization change report for the EAA Storage Reservoir Project is approximately $1.3 billion (SFWMD, 2018b). Once authorized by Congress, the revised CEPP (with EAA Storage Reservoir) could take 10 years to become operational with unconstrained funding, or 20 years with funding of $214 million/year. Table 3-4 lists the expected benefits of the project above and beyond currently authorized CERP projects.

Together with other authorized projects, the EAA Storage Reservoir is predicted to increase average annual flows into the Everglades Protection Area by 370,000 AF (SFWMD, 2018a). This flow exceeds the 323,000 AF increase projected for the original CERP (USACE and SFWMD, 1999). Also together with other authorized projects, the EAA Storage Reservoir will reduce the number of Lake Okeechobee mean monthly high-flow discharge events to the Northern Estuaries by 63 percent and the volume of Lake Okeechobee discharge by 55 percent, not including discharge events caused by local basin runoff. The original CERP goal was an approximately 80 percent reduction in damaging discharges from Lake Okeechobee to the Northern Estuaries (SFWMD, 2018a).

USACE planning guidelines require that new CERP projects be evaluated individually, with the Future Without (FWO) project condition, which assumes that all authorized CERP projects are in place, compared to the same condi-

TABLE 3-4 Hydrologic Impacts of the EAA Storage Reservoir Compared to the Future with Currently Authorized CERP Projects

	Impact on Hydrology Beyond Other Authorized Projects
Lake Okeechobee	**Minimal hydrologic change**: • Increases frequency of lake stages in the preferred stage envelope • Slightly increase in extreme high lake stages • Improves water shortage cutback performance
St. Lucie Estuary	**Moderate hydrologic change**: • Reduces mean monthly flows over 2,000 cfs and 3,000 cfs by 13%, but slightly increases mean monthly flows under 350 cfs (by 1 month) • Provides an additional 55% reduction in high-flow discharge events lasting longer than 42 days, which are particularly damaging to oysters • Reduces the number of Lake Okeechobee events that exceed the preferred salinity envelope by 39%
Caloosahatchee Estuary	**Moderate improvement**: • Reduces mean monthly flows over 2,800 cfs and 4,500 cfs by 13% and 17%, respectively, but increases mean monthly flows under 450 cfs by 12% • Provides an additional 40% reduction in high-flow discharge events lasting longer than 60 days, which are particularly damaging to oysters • Reduces the number of Lake Okeechobee events that exceed the preferred salinity envelope by 45%
Central Everglades	**Negligible to moderate improvement**: • Increases average annual flows into the Everglades Protection Area by approximately 160,000 AF • Provides moderate improvement to WCA-2A with increased stage under all conditions • Provides moderate beneficial effects on Northwest WCA-3A, reduces dryouts • Produces negligible effects on Central WCA-3A and WCA-3B • Provides minor beneficial effects on Southern WCA-3A with reduced stages during the wettest conditions • Provides minor improvements in hydrologic condition to Everglades National Park
Florida Bay	**Minor beneficial effects**: • Increases combined average annual overland flows from southern Everglades National Park to nearshore Florida Bay by 7,000 AF, providing some reduction in salinity

SOURCE: SFWMD, 2018a.

tions but with the proposed new project in place. This approach is reasonable for determining the benefits of the project alone, but it neglects the synergistic systemwide effects of future projects or possible changes in lake regulation, which could result in selection of a suboptimal plan. As mentioned previously, the EAA Storage Reservoir analyses included a model run that combined the tentatively selected EAA Reservoir plan with one early planning alternative from the Lake Okeechobee Watershed Restoration Project, consisting of 200,000 AF of surface storage and 80 5-MGD ASR wells, to provide a preliminary estimate of the cumulative impacts of these projects. This alternative represented significantly more above-ground storage than was ultimately included in the Lake Okeechobee Watershed Restoration Project tentatively selected plan. The combination was shown to achieve an 86 percent reduction in the number of high-flow estuary events, a 78 percent reduction in high estuary flows by volume from Lake Okeechobee to the Northern Estuaries, and 99 percent of the original goal for CERP flows into the Everglades (SFWMD, 2018a). This preliminary modeling exercise suggests that the original CERP flow targets may be achievable with less than the original CERP storage projects, although the analysis should be rerun with the final Lake Okeechobee Watershed plan. This analysis underscores the benefits of more inclusive systemwide modeling to understand the ultimate outcomes of all CERP projects in planning, as recommended in NASEM (2016).

The EAA Storage Reservoir analyses and projection of benefits were based on historic rainfall (1965-2005). CERP agencies should consider a new approach to identifying the best alternatives for new projects that quantitatively evaluates the robustness of hydrologic and ecologic benefits to possible future climate and sea-level rise conditions, such as Robust Decision Making (RDM) or Decision Making under Deep Uncertainty (DMDU) planning processes (Groves and Lempert, 2007; Lempert et al., 2006). Otherwise, the analyses could result in selection of a project that is vulnerable to future conditions.

Projects in Planning

Progress in project planning has important implications for the location and pace of future restoration progress, as well as the restoration benefits provided by those projects. This section describes aspects of several projects for which planning is under way—the Loxahatchee River Watershed Restoration, Lake Okeechobee Watershed Restoration, and Western Everglades Restoration projects. Although the committee examined all three projects, the committee did not rigorously review the project analysis or the selection of the final project alternatives for any of the planning efforts.

Loxahatchee River Watershed Restoration

The Loxahatchee River once drained a large expanse of coastal lagoons and undeveloped wetlands to the west, north, and south through sloughs and creek channels. Water flowed into the ocean at a natural river mouth that sometimes closed after storms, when it was blocked by shoaling of sand. In 1922 a construction project cut a deep channel 6 feet below mean low water level at the river mouth, as well as built jetties on the north and south sides of the inlet, each measuring 300 feet in length. Between 1942 and 1947, sand again blocked the inlet and since then it has been kept open with twice-yearly maintenance dredging by the Jupiter Inlet District. The jetties have been extended in length and hardened with concrete and large boulders (Figure 3-15). Deepening of the Jupiter Inlet has allowed considerably more ocean water to move upstream at high tide compared to the past. There also is less freshwater input to the river from its basin, because much of the former wetlands have been converted to farms and urban and suburban land uses (FDEP, 2010). Taken together, greater upstream movement of ocean water and reduced freshwater inputs have resulted in salinization of the river in areas that once were fresh.

The intrusion of saltwater far up the river and onto its floodplain has led to a major change in the landscape. In particular, up-river migration of mangrove has displaced cypress (FDEP, 2010). Progressing upstream by boat, one can observe dense mangrove for many miles (Figure 3-16), and in some places remnant dead cypress trees. Then cypress begins to appear, far back from the river channel in the floodplain, until one finally reaches a place where salinity is low enough that a natural cypress community exists right to the banks of the river. The Loxahatchee River Watershed Restoration Project (Figure 3-3, No. 13)

FIGURE 3-15 The deep navigation channel at the mouth of the Loxahatchee River.

SOURCE: Karl Havens.

FIGURE 3-16 A section of the Loxahatchee River where mangrove has replaced cypress, with dead cypress trees visible in the background (BOTTOM), and a section of the river further upstream with natural cypress forest to the river's edge (TOP).

SOURCE: Karl Havens.

is intended to address the issue of salinization of the river, as well as other issues in the watershed. The full suite of project objectives are as follows (USACE and SFWMD, 2018a):

- Improve water distribution and timing to restore the natural system's ecological functioning,
- Rehydrate natural areas that have been hydrologically impacted by excessive draining and water diversion,
- Reestablish connections among natural areas that have become spatially and/or hydrologically fragmented,
- Improve timing and distribution of water from the upstream watershed to increase the resiliency of freshwater riverine habitats to future sea-level changes, and
- Restore, sustain, and reconnect the area's wetlands and watersheds that form the historic headwaters for the river.

The Project Delivery Team is evaluating a rescoped project, and a report describing the results was expected in September 2018, which was too late for the committee's review. The team is evaluating the performance of four alternatives that combine different water storage and conveyance features (Table 3-3). These include different amounts of deep and shallow storage in the L-8 basin, different amounts and types (constructed vs. natural) of storage in the C-18W basin, different numbers of aquifer storage and recovery (ASR) wells, and different approaches to flowing water through three delivery routes in the headwaters of the river. The alternatives involve a complex array of interconnected projects

TABLE 3-3 Loxahatchee River Watershed Restoration Project Alternatives

Alternative	Deep Storage in L-8 Basin	Shallow Storage in L-8 Basin	C-18W Basin Storage	Aquifer Storage and Recovery (ASR)	Primary Delivery Route	Secondary Delivery Route	FW 3 Features
2	None	4,300 ac-ft	7,200 ac-ft reservoir	2 wells at C-18W storage	FW2	FW1	Full range
5	None	None	9,500 ac-ft reservoir	4 wells at C-18W storage	FW2	FW1	Full range
10	44,000 ac-ft	None	7,200 ac-ft reservoir	None	FW1/FW2		Limited
13	None	6,500 ac-ft	Increased wetland elevations to support natural storage	4 wells at L-8 storage	FW2	FW1	Full range

SOURCE: Foster, 2018.

that include adding spreader canals, plugging ditches, regrading land, replacing weirs, and constructing reservoirs, pumps, and/or ASR wells. An example is shown in Figure 3-17. Although the evaluation of alternatives is ongoing, early results indicate potential benefits for the river salinity. Results from a hydrodynamic model indicate that salinities can be substantively improved (Figure 3-18).

A major driver of this project implies that cypress habitat is of greater value than mangrove habitat, even though both are native Florida plant assemblages with ecosystem values. Further, cypress requires a particular inundation regime, not just a certain salinity, and it is not clear whether this was considered dur-

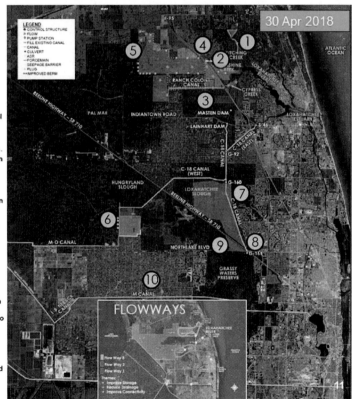

FIGURE 3-17 The construction features of Alternative 5 of the Loxahatchee River Watershed Restoration Project.

SOURCE: Foster, 2018.

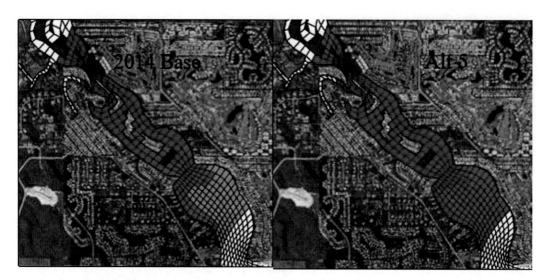

FIGURE 3-18 Results of a hydrodynamic model simulation of the percentage of time that salinities fall within a desired range to support halting the advance of mangrove up the main stem of the Loxahatchee River.

NOTE: The comparison illustrates a potential for improvement—here between Alternative 5 and the no action base based on increasing freshwater flows. The colors correspond to percentages of time that salinities are in the desired range. The light blue color is <20%, deep blue 20-40%, dark blue 40-60%, and purple 60-80%. The region that most often meets the salinity performance target is pushed downstream, and an area develops up-river with salinities meeting the target more than 60% of the time, which does not exist in the base.

SOURCE: SFMWD, 2017.

ing project development or evaluation. Finally, of the various CERP restoration projects, this project is among those particularly sensitive to the effects of sea-level rise, which was not incorporated explicitly into alternatives analysis and selection (J. Leeds, SFWMD, personal communication, 2018). The agencies assume that they will be able to overcome this issue through adaptive management, which becomes a greater factor in affecting salinity in the mangrove-cypress transition area every year. Yet it remains to be seen whether any amount of adaptive management can be effective out to 2050, especially in times of drought. The CERP agencies have mentioned plans to examine potential effects of sea-level rise on project benefits, which the committee considers to be a critical next step because future sea-level rise could potentially negate this project's value, at least regarding prevention of further upstream expansion of mangrove.

Lake Okeechobee Watershed Restoration Project

The Lake Okeechobee Watershed Restoration Project located north of Lake Okeechobee is intended to provide additional storage to improve the quantity, timing, and distribution of flows into Lake Okeechobee to reduce the occurrence of extreme high and low water levels, improve systemwide operational flexibility, reduce high-volume estuary discharges, increase the spatial extent and condition of wetland habitat, and increase available water supply (M. Ferree, SFWMD, personal communication, 2018). Project goals do not include improving delivery of water to the remnant Everglades. The 3-year USACE planning process began in 2016. Project features that are being evaluated to deliver benefits include reservoirs, ASR wells, wetland/floodplain restoration, perpetual flowage easements, and interbasin transfers. Deep injection wells to dispose of water were discussed in early planning meetings but are not part of the proposed project.

Based on technical analyses, numerical modeling, public input, and other factors, the project delivery team identified three alternatives, each including reservoir storage (ranging from 65,000 to 276,000 AF), ASR wells, and wetlands restoration on the north side of Lake Okeechobee. From this analysis, the team tentatively selected a plan that includes 43,000 AF of storage in a shallow (4 feet average depth) "wetland attenuation feature" designed to enhance habitat benefits, 80 ASR wells (with 448,000 AF/year storage capacity), and approximately 5,300 acres in wetland restoration (Figure 3-19).

The $1.4 billion tentatively selected plan is predicted to reduce the number of months of high water discharge (>2,800 cfs) from Lake Okeechobee to the Caloosahatchee Estuary by 10 percent (or by 7 months in the 41-year modeled scenario) over the future without project scenario. The plan is predicted to reduce the number of 2-week periods of high discharge from Lake Okeechobee by 46 percent in the 41-year simulation (or 17 times that the 14-day moving average is greater than 2000 cfs for more than 14 days). The tentatively selected plan is predicted to increase the percentage of time that the lake is in the ecologically preferred stage envelope from 28 to 32 percent and to reduce water supply cutbacks by 33 percent. The full analysis is documented in the project implementation report, which was released in July 2018 (USACE and SFWMD, 2018c). Because this date was late in the committee's study process, the committee was unable to review the analysis in detail.

The specific benefits provided by ASR relative to those provided by shallow storage or by revisions to Lake Okeechobee operations were not identified. Because ASR provides 448,000 AF/year in subsurface storage compared to 43,000 AF of dynamic storage in the shallow storage feature, some critical uncertainties remain to be resolved with large-scale ASR before the project is implemented. NRC (2015) highlighted concerns regarding ecotoxicology that

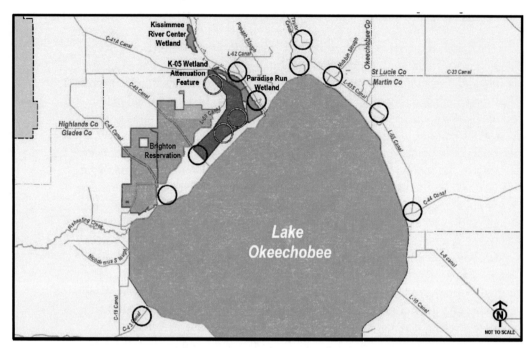

FIGURE 3-19 Features of the Lake Okeechobee Watershed tentatively selected plan as of May 2018.

SOURCE: USACE and SFWMD, 2018b.

were not resolved in the ASR Regional Study (USACE and SFWMD, 2015b) and suggested further field-scale research to resolve these questions, possibly using a cluster of ASR wells to examine the effects. Such testing would need to be incorporated into the design plan.

The modeling of project results does not account for future precipitation and temperature changes. The resiliency of each alternative to changing environmental conditions associated with climate and sea-level rise should be evaluated for all projects, but especially for one that considers the effects of lake stage on habitat conditions. Precipitation scenarios should be examined to determine whether some alternatives provide greater resilience under a wider range of future scenarios. In addition, the analyses should consider the benefits south of Lake Okeechobee to inform understanding of the systemwide benefits, avoid uncertainty regarding the ultimate disposition of water that is not discharged to the estuaries as a result of the project, and enable review of the results in the context of other planned projects, such as Central Everglades Restoration and the EAA Storage Reservoir.

Although the USACE does not consider unauthorized projects when projecting benefits, in this case the process would have benefited from analysis of the systemwide effects of the tentatively selected plan for the Lake Okeechobee Watershed Restoration Project and the proposed EAA Storage Reservoir Project in combination. The committee is concerned that the project-by-project analysis conducted thus far may result in selection of suboptimal alternatives and skew the estimation of environmental benefits attributable to individual projects. In early 2018, before the tentatively selected plan was finalized, an analysis of the combined projects assumed much larger above-ground storage for the Lake Okeechobee Watershed Restoration Project (200,000 AF) than what was ultimately chosen. Although an important first step toward more integrated analyses, this analysis should be repeated with the final plan. Modeling of the system at a larger geographic extent would also enable evaluation of synergies that would otherwise not be observable (see also Chapter 6).

Western Everglades Restoration Project

The term "western Everglades" is commonly used to refer to the Everglades landscape that extends westward from WCA-3A and the Everglades Agricultural Area and encompasses Big Cypress National Preserve, as well as reservations of the Seminole and Miccosukee tribes (Figure 3-20). This area suffers from impaired water quality, particularly from phosphorus-laden runoff from agriculture landscapes in the north and altered hydrology. Elevated nutrient levels have spurred changes in flora and fauna, degrading the biodiversity of the region and impacting habitats used for traditional cultural practices. Unnaturally high water levels drown tree islands and inundate nesting habitat along the perimeter of WCA-3A, while unnaturally dry conditions promote wildfires elsewhere within the western Everglades. The Western Everglades Restoration Project (WERP) is intended to reestablish ecological connectivity; restore hydroperiods and pre-drainage distributions of sheet flow; restore low-nutrient conditions to reestablish native vegetation; and promote ecosystem resilience. At 1,200 square miles, the WERP footprint is large, covering an area equivalent to the CEPP.

The WERP planning process is ongoing, and selection of a tentative plan is anticipated in early 2019. The three alternative plans under initial consideration shared several features, including new water control structures (weirs, culverts), levee gaps, and canal backfills, that were intended to improve hydration patterns within the interior portions of Big Cypress National Preserve and Seminole Big Cypress Indian Reservation. Each plan included stormwater treatment facilities north of the tribal areas designed to reduce nutrient concentrations associated with new water flows to meet downstream water-quality standards. While Alter-

78   *Progress Toward Restoring the Everglades*

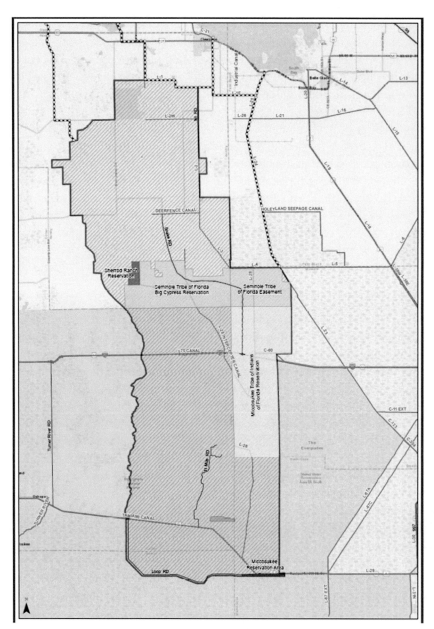

FIGURE 3-20 The officially defined geographic extent of the western Everglades for the Western Everglades Restoration Project.

SOURCE: http://www.saj.usace.army.mil/Missions/Environmental/Ecosystem-Restoration/Western-Everglades-Restoration-Project/.

native 1 relied on a rain-driven, passive management approach, Alternatives 2 and 3 incorporated more active management features, including ASR wells and a new connection for delivering water from Lake Okeechobee. Alternative 3 also incorporated new canals that connect northern parts of the project area with Lake Okeechobee, the Caloosahatchee River, and the Miami Canal, which could provide some additional relief to the northern estuaries that receive flows from the lake.

The planning team evaluated these three alternative plans using regional and subregional models, and, as of May 2018, is developing two revised alternatives for a second round of modeling based on the features from the original plans that performed well. The planning team has also developed a resilience performance measure to assess the performance of the project in terms of its capacity to sustain ecological benefits under conditions of stress. The approach will involve training a stochastic hydrologic emulator tool, as described by Ali (2009), to approximate the hydrologic responses computed by the Regional Systems Model (RSM) for a historical period of record (e.g., 1965-2005). The trained emulator will simulate project outcomes, but with synthetic rainfall and temperature records that include periods of stress, such as a prolonged drought or a wet season with especially intense rainfall. The approach is still in development and was originally intended to be applied to all alternatives but is currently (as of October 2018) anticipated to be applied only to the tentatively selected plan (D. Crawford, USACE, personal communication, 2018). The committee lauds this effort to evaluate project resilience in the context of an uncertain future that may bring more frequent extreme events capable of threatening CERP infrastructure and of imperiling recovered and recovering portions of the ecosystem. However, analysis of resilience would be even more useful if applied earlier in the analysis of alternatives.

CERP Pilot Projects: The Decomp Physical Model

Pilot projects are limited efforts designed to provide scientific or engineering knowledge that can be applied to improve major restoration projects. They provide the opportunity to experiment with methods and approaches without incurring the large expense of fully developed restoration projects. Only one CERP pilot project is currently under way—the Decomp Physical Model (DPM). The DPM is a large-scale active adaptive management project to improve understanding of how degraded portions of the ridge and slough landscape might respond to increased water deliveries and to inform decision making regarding restoration project design and operational targets for flow in the ridge and slough landscape (see project details in Box 3-1). Four flow experiments were

BOX 3-1
DPM Project Design

The DPM experiment was conducted between L-67A and L-67C, in an area near the border of WCA-3A and WCA-3B known as the "the pocket" (see Figure 3-1-1). The project components included 10 gated culverts on the L-67A canal (referred to as S-152) and a 3,000-foot gap created in the L-67C levee with three back-fill "treatments" in the adjacent canal. The canal was left completely open for the northern-most treatment, while the center and southern-most treatments had partial and complete backfills, respectively (Figure 3-1-1). Data were collected prior to experimentation as well in four periods following construction using a before-after control-impact (BACI) structured design. A 2-month flow experiment was initiated in November 2013 by opening the 10 gated culverts that comprise S-152 (Figure 3-1-1). Subsequent flow experiments were conducted for 3 months starting in November 2014, 2 months starting in November 2015, and 3 months starting in October 2016.

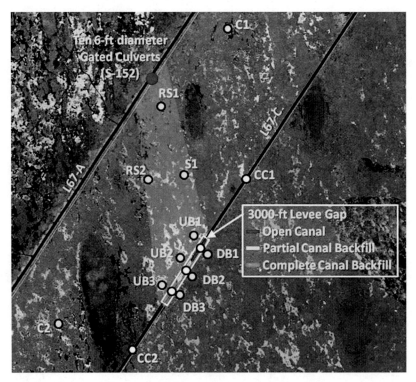

FIGURE 3-1-1 Map of the Decomp Physical Model located in "the pocket" between L-67A and L-67C.
SOURCE: Sklar, 2013.

executed between 2013 and 2016, and NASEM (2016) detailed the scientific findings through mid-2016. In the fourth year of the pilot, landscape modifications and sawgrass removal were explored to evaluate the potential of active slough management to reshape flow characteristics and to increase sediment movement, thereby serving as a way to jump-start restoration.

The DPM is entering a second phase of testing that will occur from 2018 to 2021. The second phase will allow for year-round discharges from S-152 subject to operational constraints based on phosphorus levels in L-67A and stages in WCA-3B and nearby canals. It will also include a large-scale active management component to increase sheet flows by modifying slough structure within the pocket. Key questions guiding the second phase of the DPM are summarized in Box 3-2. Answers to these questions will reduce uncertainties related to the spatial and temporal scales of sheet flow reestablishment and biological responses to the restored flows. Interest is especially focused on the spatial extent of restoration of sloughs and the velocities required for restoration. By extending the flow from a few months to year round, hydrologic connectivity of currently fragmented sloughs may increase. Finally, research on backfilling of canals will provide quantitative insights into sediment and phosphorus loading in the canals and the potential effects on downstream marsh areas (Choi et al., 2017). No DPM phase 2 results were available to review as of June 2018.

BOX 3-2
Key Questions of the DECOMP Physical Model Phase 2

Sheet Flow
- Changes in slough periphyton will result from changes in sheet flow. Over what time period and distance will these changes occur?
- What are the food-web responses to increased flows and phosphorus loading?
- How effective is the S-152 in restoring large areas? Will sheet flow only restore small areas (500 m radius from S-152), or will impacts spread over larger areas when active management is used?
- Will longer sheet flow duration increase hydrologic connectivity between the Pocket and WCA-3B?

Canal Backfill
- Will phosphorus-enriched sediments in canals that are mobilized by higher flows have an impact on areas downstream?
- How will submerged and emergent vegetation change and stabilize over a longer period in the partial and complete backfilled canal sections?

SOURCE: Choi et al., 2017.

NON-CERP RESTORATION PROGRESS

CERP projects are not the only restoration efforts ongoing in the Everglades region. Several non-CERP projects are critical to the overall success of the restoration program, and their progress directly affects CERP progress. This section reviews new information on major non-CERP efforts, with an emphasis on natural system restoration benefits or implications for CERP progress. Progress in non-CERP projects that contribute to restoring flow, improving water quality, and controlling invasive species are also discussed.

Restoring Flow to Northeast Shark River Slough and Taylor Slough

Several non-CERP projects contribute to the overall objective of increasing flows in Northeast Shark River Slough and Taylor Slough within Everglades National Park. These include the Limestone Products Association Seepage Management project, the Modified Water Deliveries to Everglades National Park Project (Mod Waters), the C-111 South Dade Project, and the Florida Bay Initiative (Figure 3-21). Several of these projects, now completed or nearing completion, will operate together through the Combined Operations Plan to maximize the ecosystem benefits of the new water control features.

Limestone Products Association Seepage Management

Seepage management involves regulating the exchange of groundwater from natural areas into developed areas, which are separated from one another by canals, levees, or other structures. During the wet season in particular, the L-31N Canal diverts groundwater, drawn primarily from the northeastern portion of Everglades National Park, to the C-111 basin in south Miami-Dade County. As part of a non-CERP project constructed in exchange for wetland mitigation credits, the Miami-Dade Limestone Products Association constructed a seepage barrier in this area to reduce this groundwater discharge to the L-31N Canal, thereby increasing water levels and promoting greater sheet flow in northeast Shark River Slough.

Phase 1 of this project began in 2012 with the construction of a 2-mile-long seepage barrier as a pilot project. Construction of the barrier involved excavating a 32-inch-wide trench to a depth of 35 feet below the ground's surface. The trench was filled with a concrete-bentonite slurry formulated specifically for this application. Phase 2 of the project—a 3-mile extension of the seepage barrier—was constructed in 2016 (see Figure 3-21).[8]

---

[8] See http://www.l31nseepage.org/index1.html.

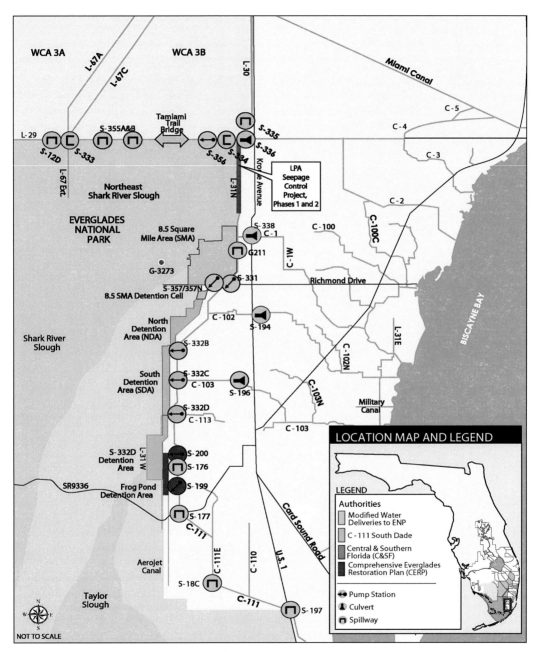

FIGURE 3-21 Mod Waters, C-111 South Dade, and the Limestone Product Association seepage barrier are all expected to contribute to increased flows in Northeast Shark River Slough and Taylor Slough in Everglades National Park.

SOURCE: Modified from USACE, 2017, fact sheet.

Hydrologic measurements demonstrate the barrier is reducing seepage into the canal. Recent monitoring data indicate that the 3-mile Phase 2 barrier extension increased groundwater elevation differences across the barrier by about 0.7 to 1.2 feet during the wet conditions of 2016 and 2017, which indicate greater resistance to groundwater flow (Figures 3-22 and 3-23). This results in additional hydraulic head within the wetland to drive sheet flow through Shark River Slough. The available data are encouraging and suggest that the project is satisfying its objectives.

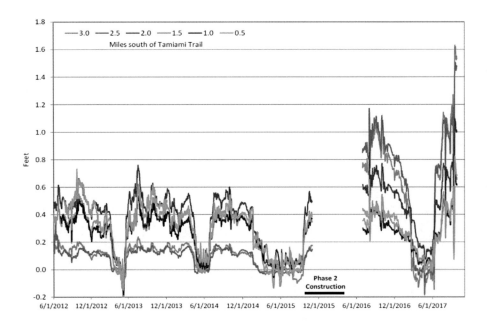

FIGURE 3-22 Difference in groundwater levels east and west of the Limestone Products Association seepage barrier at points 0.5 to 3.0 miles south of Tamiami Trail.

NOTE: Phase 1 of the seepage barrier extended 2 miles south of Tamiami Trail, so the 2.5 and 3.0 mile monitoring stations (orange and blue lines) served as control sites, unaffected by the project through late 2015. Phase 2 of the project extended the barrier another 3 miles to the south, and the post-construction monitoring shows even greater differences in water levels for the 2.5- and 3-mile sites than the Phase 1 sites. These central sites are the farthest from the ends of the seepage barrier.

SOURCE: T. MacVicar, MacVicar Consulting, personal communication, 2017.

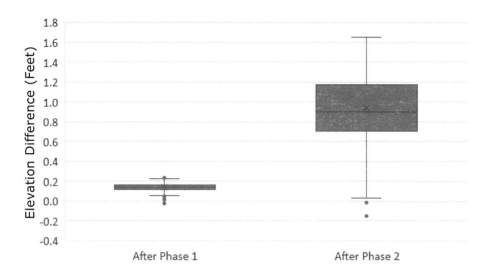

FIGURE 3-23 Difference in wet season groundwater elevations at 3 miles south of Tamiami Trail, which is south of the 2-mile Phase 1 barrier but near the center of the barrier after the 3-mile Phase 2 addition.

NOTE: "After Phase 1" includes wet season data from 2012 to 2015, and "After Phase 2" includes wet season data from 2016 to 2018.

SOURCE: T. MacVicar, MacVicar Consulting, personal communication, 2018.

Modified Water Deliveries to Everglades National Park Project

Congress provided legislative authority in 1989 for the creation of a project to improve water flows into Everglades National Park, where Everglades microtopography and vegetation were in decline as a result of insufficient inflows. In 1992, the General Design Memorandum (GDM) for the Modified Water Deliveries to Everglades National Park Project (Mod Waters; USACE, 1992) envisioned several features to increase the flow of water from WCA-3 into Everglades National Park to accommodate flows up to 4,000 cfs. Increasing the flow of water from WCA-3A into Northeast Shark River Slough is a central aspect of Everglades restoration, and the capacity for successful southward movement of waters provided by the CEPP and other future CERP projects depend critically upon the conveyance, seepage reduction, and flood management provided by Mod Waters. Hence, completion of Mod Waters is essential to the ultimate success of the CERP. An extensive discussion of the history and details of this long-delayed project was provided by NRC (2008), and updates on progress are provided in each of the succeeding biennial reviews.

Construction of all major components of Mod Waters was completed in May 2018, including flood mitigation in the 8.5-square-mile area, conveyance and seepage control, and the 1-mile bridge on Tamiami Trail (see Figure 3-21; Gonzales, 2018). Operation of the project will be determined based on the Combined Operational Plan, discussed later in this section.

Although the 1-mile Tamiami Trail bridge is the sole bridging feature within Mod Waters, a 2.3-mile western bridge is under construction under the Tamiami Trail Modifications: Next Steps Project, with construction completion estimated in December 2018. Engineering design is also under way on Tamiami Trail Next Steps Phase 2, which would include raising 6.7 miles of roadbed to accommodate CERP design water levels of 9.7 feet in the L-29 Canal and installing additional culverts and concrete arches in the locations of historic sloughs (R. Johnson, DOI, personal communication, 2018). These projects combined with construction of the Central Everglades S333-N gated spillway (estimated to start construction in October 2018) are projected to improve the capacity to move water from WCA-3A into Everglades National Park, reducing the damage associated with high water conditions in the WCAs and increasing flows to Northeast Shark River Slough.

C-111 South Dade

The C-111 South Dade Project provides the connection between Mod Waters and L-31N Seepage Management projects to the north and the C-111 Spreader Canal (Western) Project to the south (described earlier in this chapter; see Figure 3-21). This major modification to the Central and South Florida (C&SF) Project's C-111 Canal was authorized in 1994 to maintain existing flood protection and other C&SF project purposes in developed areas east of C-111, while restoring natural hydrologic conditions in the Taylor Slough and eastern panhandle areas of Everglades National Park (USACE, 2015b). Increased freshwater flows in these areas also help conditions in Florida Bay.

The C-111 South Dade Project consists of a combination of large detention areas and levees, pump stations and structures, bridges, and backfilling (Figure 3-21; USACE, 2015b). The overall project contributes to maintenance of the hydraulic ridge along the C-111 corridor, thereby reducing seepage to the east. The South Detention Area was completed and operational in 2010. Benefits from the project operations to date are discussed collectively with the C-111 Spreader Canal Project, discussed earlier in the chapter. The final features of the C-111 South Dade Project are scheduled for completion by fall 2018 (Gonzales,

2018). The C-111 South Dade Project will ultimately be operated and evaluated as part of the Combined Operational Plan.

Combined Operational Plan

As the final components of the Mod Waters and C-111 South Dade projects are nearing completion, efforts have recently been focused on developing the Combined Operational Plan for the system that will be used to meet the overarching objectives of the projects. These objectives include increasing flows from WCA-3A into Northeast Shark River Slough, maintaining higher water levels in Everglades National Park without exacerbating flooding in suburban and agricultural lands to the developed east, increasing flows to Taylor Slough and Florida Bay, and reducing regulatory discharges from WCA-3A through the S-12 structures or south through the South Dade Conveyance Canals. In 2015, in cooperation with Everglades National Park, the USACE began a phased implementation of operations of Mod Waters and C-111 facilities to move more water into Northeast Shark River Slough and obtain data needed to develop the operating plan (Box 3-3; NPS, 2016; USACE, 2016b). The Everglades Restoration Transition Plan (ERTP; George, 2016) previously defined operations for the constructed features of the Mod Waters and C-111 South Dade projects. Completion of the Combined Operational Plan is anticipated in 2020.

Florida Bay Project

In July 2016, the SFWMD launched a new state-funded initiative to move more water into Taylor Slough and Florida Bay via infrastructure modifications in the area near the S-332D detention area and the Frog Pond (see Figure 3-21). Modifications include expedited completion by the SFWMD of authorized C-111 South Dade features, which included adding 10 plugs, an additional weir in the L-31W Canal, and levee removal in the S-332D detention area to promote more flow into Taylor Slough. Other work includes sealing the S-332D discharge basin and adding pump capacity in the S-199 and S-200 pumps in the C-111 canal (W. Wilcox, SFWMD, personal communication, 2018). A new connection between the Frog Pond detention area and the L-31W canal (G-737; see Figure 3-24) was added by the SFWMD under the Florida Bay Project and completed in 2017 (Qui et al., 2018).

Rather than constructing a hydraulic ridge as is used in the C-111 South Dade and C-111 Spreader Canal Western projects, the Florida Bay Project enhances direct discharge into Taylor Slough. The project initially raised concerns about whether additional loads of phosphorus associated with higher

> **BOX 3-3**
> **Phased Implementation of the Combined Operational Plan**
>
> During the three-phase implementation of the Combined Operational Plan, field testing and monitoring is being used to evaluate the hydrologic response to the proposed new operations.
>
> - **Increment 1** (2015-2017) relaxes existing constraints on gage G-3273 related to flow from WCA-3A into Northeast Shark River Slough, while maintaining the L-29 Canal at the stage of 7.5 feet National Geodetic Vertical Datum (NGVD). Increment 1 is also designed to test seepage control provided by the S-356 pump station, which was designed to return seepage water back into Northeast Shark River Slough from the L-31N Canal. Increment 1 testing also includes reduced flow to South Dade from S-331 and conditional increased use of S-197 (USACE, 2015a). Increment 1 was initiated in October 2015, but was interrupted from December 2015 to December 2016 by flood management operations and an emergency deviation due to high water levels in WCA-3A and the time required for recovery. During this deviation, water levels in the L-29 Canal exceeded 8 feet NGVD for more than 2 months.
> - **Increment 1 Plus** (2017-2018) is an update to Increment 1 and incorporates the lessons learned from the emergency deviation and the Reasonable and Prudent Alternative (RPA) from the July 2016 Everglades Restoration Transition Plan Biological Opinion (FWS, 2016). Increment 1.1, implemented in March 2017, incorporates these changes while maintaining an L-29 Canal stage of 7.5 feet NGVD. Increment 1.2 will raise the L-29 Canal stage to 7.8 feet NGVD, pending completion of C-111 South Dade Contract 8 flood management features, anticipated in August 2018. A temporary deviation was implemented between June 28, 2017, and October 2017 to maximize high water discharges out of the WCAs.
> - **Increment 2** (2018-2019) was approved in February 2018 and implemented in July 2018. Increment 2, which allows the L-29 Canal to reach a maximum stage of 8.5 feet, further relaxes constraints set by G-3273 and tests seepage control from the S-356 pump station (USACE, 2018f; R. Johnson, NPS, personal communication, 2018).
> - In **Increment 3** (2019) the new combined operational plan for the system will be developed using data collected during Increments 1 and 2.

flows of water that meets all water quality criteria would trigger an ecological imbalance in Taylor Slough. State and federal agencies are partnering to support a comprehensive monitoring and adaptive management plan to examine the project effects on downstream water quality, floc nutrient content, periphyton, vegetation, invertebrates, and fish in Taylor Slough. The committee supports this initiative and the comprehensive monitoring and adaptive management program. This effort should greatly inform understanding of the ecosystem response to increases in nutrient loads associated with increases in discharge and should resolve uncertainties in a structured process along the way.

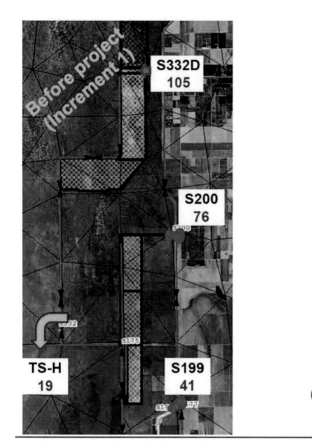 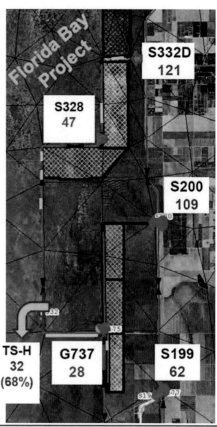

FIGURE 3-24 Modeled average annual flows associated with the Florida Bay Project show increasing flows (in kAF) in red toward Taylor Slough from several structures, including a 68 percent increase in discharge into the headwaters of Taylor Slough.

SOURCE: W. Wilcox, SFWMD, personal communication, 2018.

Everglades Water Quality Initiatives

The Everglades is an oligotrophic system whose productivity is limited by phosphorus. Historically, concentrations of phosphorus were very low in soil and water. Elevated inputs of phosphorus from agriculture, urban development, and other human activities can and have altered the structure and function of the wetland. The Everglades Forever Act and the water quality standards for phosphorus in the Everglades Protection Area established the concentrations of total phosphorus necessary to protect the ecosystem. These values are used to guide water quality management in the Everglades using field data from monitoring programs. Achieving these water quality goals is critical to progress in moving water into the Everglades Protection Area (see Chapter 2), and therefore, progress addressing water quality throughout the watershed has implications for CERP progress. Phosphorus dynamics in Lake Okeechobee, which will be the source of most of the new water in the CERP, is discussed in Chapter 4.

Stormwater Treatment Areas

The state of Florida's 1994 Everglades Forever Act authorized the construction and operation of large constructed wetlands, known as STAs. These STAs are central to the state efforts to restore and protect water quality and maintain the structure and function of the Everglades Protection Area. In 2012, the SFWMD developed its "Restoration Strategies" plan, which provides for expanding existing STA acreage to meet the water quality–based effluent limitation (WQBEL) for discharges from the STAs. The WQBEL requires (1) a maximum total phosphorus annual flow-weighted mean of 19 parts per billion (ppb) and (2) a long-term flow-weighted mean of total phosphorus concentration of 13 ppb not to be exceeded in more than 3 of 5 years (Leeds, 2014).

The STAs are designed and operated by the SFWMD to decrease total phosphorus concentrations in surface water prior to discharge into the Everglades Protection Area. Including the recent expansions of STA-2 and STA-5/6, pursuant to the SFWMD's Restoration Strategies, a total of 57,000 acres of treatment area is currently permitted to operate. To support the operation of STA-3/4 and STA-2, the A-1 FEB, with 56,000 AF of water storage capacity, was completed in 2016. A second FEB (L-8) with approximately 45,000 AF of water storage capacity to improve the function of STA-1E and STA-1W should be completed in 2018 (see Figure 3-25 and Table 3-5). Flow equalization basins should improve STA performance by limiting large water and phosphorus inputs during wet periods to ensure adequate treatment and supplying water during dry periods to ensure that treatment cells retain some water to maintain their function.

For the operational period from 1995 to present, the STAs in total have treated approximately 18.6 million AF of water (~6 trillion gallons). This treatment has removed 2,300 metric tons (mt) of total phosphorus from water flows entering the Everglades Protection Area, with an overall removal rate of 77 percent (Figure 3-26). During this period, the flow-weighted mean concentration of total phosphorus in the outflow has been 31 ppb.

In 2017, the STAs treated a combined 1.1 million AF of water, retaining 108 mt of total phosphorus. This treatment level is equivalent to an 84 percent reduction in the total phosphorus load, resulting in a flow-weighted mean of 15 ppb

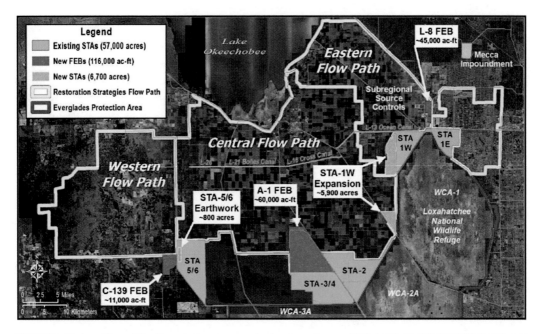

FIGURE 3-25 Location of the Everglades stormwater treatment areas (STAs) STA-1E, STA-1W, STA-2, STA-3/4, and STA-5/6 and the planned locations for additional STAs, STA earthwork, and flow equalization basins (FEBs) associated with the Restoration Strategies plan.

SOURCE: SFWMD.

total phosphorus concentration in the outlet water. This level is the lowest flow-weighted mean concentration of total phosphorus attained to date for the combined annual discharge from the STAs. The flow-weighted mean concentrations of total phosphorus in outflow of individual STAs ranged from 11 ppb (STA-3/4) to 23 ppb (STA-1W) in 2017, while the percentage of the total phosphorus load retained in the STAs ranged from 80 (STA-3/4) to 87 percent (STA-1E, STA-5/6).

Since the early 2000s, the STAs have steadily improved in phosphorus removal from water inflows, despite marked event-based, seasonal, and annual variation in water and phosphorus loads. With the FEBs coming on line, the performance of the STAs is expected to continue to improve. The performance of the STAs is approaching the WQBEL discharge limit. If the removal of total phosphorus continues to improve and remains below the discharge limit of 13 ppb over the longer term (more than 2 of 5 years), then STA discharges can be redistributed as contemplated by the CERP and the CEPP.

TABLE 3-5 Summary Status of Major Restoration Strategies Project Elements

Component	Purpose	Status	Construction Completion
Eastern Flowpath			
L-8 FEB	Attenuate flow into STA-1E and -1W	Completed	June 2017
L-8 Conveyance Features (G-716, G-341, G-541)	Assist movement of inflows and outflows to L-8 FEB	Most completed; G-341 under construction	TBD; final phase to begin February 2019
STA-1W expansion (Phase 1)	Increase STA-1W effective treatment area	Under construction	December 2018
STA-1W expansion (Phase 2)	Increase STA-1W effective treatment area	Design under way	TBD
Central Flowpath			
A-1 FEB	Attenuate flow into STA- 2 and -3/4	Completed	2015
Western Flowpath			
STA 5/6 Earthwork	Improve the performance of STA-5/6	Under construction	TBD
C-139 FEB	Attenuate flow into STA- 5/6	Design underway	TBD

SOURCE: https://www.sfwmd.gov/documents-by-tag/resstrategies; Jacoby, 2018; M. Jacoby, SFWMD, personal communication, 2018.

Water Quality in the Everglades Protection Area

The SFWMD has developed mass balances of water and nutrients for the Everglades. These mass balances are useful because they provide a systemwide perspective on the inputs and processing of important constituents (Figure 3-27) (Julian et al., 2018). Figure 3-27 shows that over the most recent 5-year period (2013-2017), 139 mt of total phosphorus was exported from Lake Okeechobee to the EAA or STAs. After flow through the EAA, 193 mt entered the STAs, with only 34 mt exported from the STAs.

Time series analysis of the period of record (1979-2016) has shown significant decreases in annual geometric mean concentrations of total phosphorus for inflows to the Loxahatchee Natural Wildlife Refuge (WCA-1), WCA-2, and WCA-3, which reflects the continued improvement of the phosphorus removal of the STAs. Similar trends were also evident over the more recent 2005-2016 time period. Total phosphorus concentrations in inflows to Everglades National Park have been variable and, unlike trends observed in other regions, have not steadily declined over the period of record. However, recent decreases in total phosphorus concentrations were evident for interior monitoring sites in

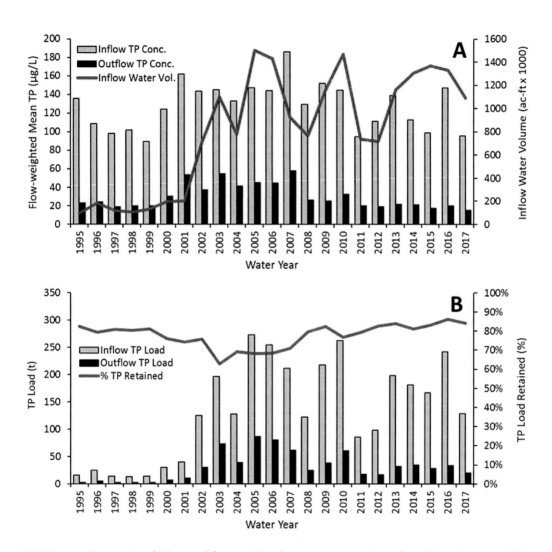

FIGURE 3-26 Time series of (A) annual flow-weighted mean concentrations of total phosphorus in inflow and outflow waters, with corresponding inflow water volumes and (B) annual inflow and outflow loads of total phosphourus with percent total phosphorus retained for the period of record of the combined STAs.

SOURCE: Chimney et al., 2018.

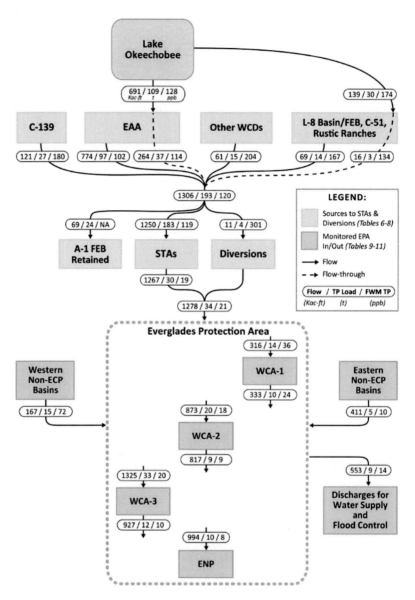

FIGURE 3-27 Five-year (2013-2017) average annual flows, total phosphorus loads, and flow-weighted mean total phosphorus concentrations from Lake Okeechobee to the EAA and STAs and across the EPA.

NOTE: WCD = water control district and ECP = Everglades Construction Project.

SOURCE: Julian et al., 2018.

Everglades National Park as well as WCA-2 and WCA-3. Seventy-one percent of the interior marsh sites had annual geometric mean concentrations of total phosphorus of 10 ppb or less, while 79 percent throughout the greater ambient monitoring network had annual geometric mean concentrations of total phosphorus of 15 ppb or less (Figure 3-28; Julian et al., 2018).

Analysis of data from the water quality monitoring program for 2013-2017 showed that unimpacted portions of the WCAs passed all four parts of the compliance test.[9] Impacted (i.e., phosphorus-enriched) sites of the WCAs, however, routinely failed one or more parts of the compliance test. Total phosphorus concentrations at some of the impacted sites were below the long-term and annual limits and appear to be transitioning from impacted to unimpacted conditions. Overall, the STAs have been effective in removing phosphorus from inflowing waters, and removal efficiency is improving with time. The Everglades water quality is expected to continue to improve as additional FEBs and STAs come on line through Restoration Strategies.

Planning for Invasive Species Management

Invasive exotic species present a major challenge for Everglades restoration (NRC, 2014). At least 192 exotic animal species are established in the Greater Everglades, and 75 plant species are listed as priorities for control by the SFWMD (NRC, 2014). Ideally, the restored Everglades would have only the species found there before European occupation, but that goal cannot be achieved. Most established exotic species in the region cannot be exterminated, and populations of some of them cannot even be controlled. The "restored Everglades" will inevitably have a combination of its pre-European biota and many exotic species. Being able to predict which exotic species are most likely to become established and cause undesirable effects would be very helpful to managers and decision makers (NRC, 2014). Perfect prediction is impossible, but partial predictive success can be extremely valuable for informing decisions concerning how to target efforts to prevent introduction of potentially invasive species and how to control already established exotic species. NRC (2014) recommended, among other things, development of a "strategic early detection and rapid response (EDRR) system that addresses all areas, habitats, and species."

A screening tool for managers to use to assess the need to initiate a rapid response, as well as the resources and knowledge available to support

---

[9] The four-part test is used to assess compliance according to the following four provisions: (1) 5-year geometric mean is less than or equal to 10 ppb, (2) annual geometric mean averaged across all stations is less than or equal to 11 ppb, (3) annual geometric mean averaged across all stations is less than or equal to 10 ppb for 3 of 5 years, and (4) annual geometric mean at individual stations is less than or equal to 15 ppb (FAC §§ 62.302.540).

96 Progress Toward Restoring the Everglades

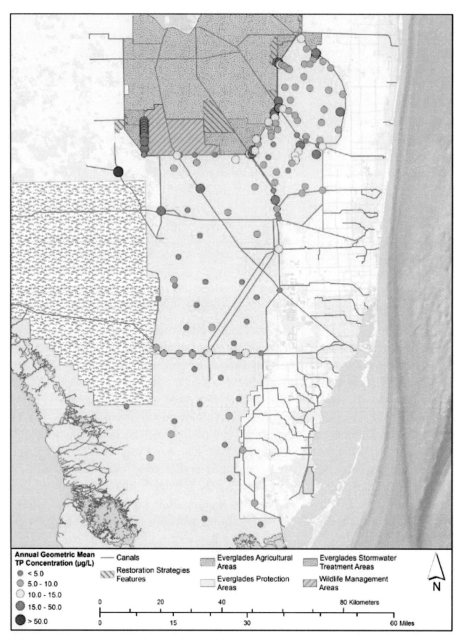

FIGURE 3-28 Annual geometric mean total phosphorus concentrations at sampling sites in and near the Everglades Protection Area in water year 2017.

SOURCE: Julian et al., 2018.

that response, is in development. The tool consists of a systematic evaluation of information about species—both potential invaders and invasive species already present—that then permits a qualitative evaluation of the likelihood that a species will become established and invasive. The tool is being tested against invasive species that exist in the system to determine how well it would have worked if it had been used with the information available at that time. To date, a retrospective analysis of more than 100 species has largely confirmed the tool's usefulness (C. Romagosa, personal communication, 2018). In addition, the tool could help guide research of established exotic species by evaluating the potential value of different kinds of information.

CONCLUSIONS AND RECOMMENDATIONS

CERP project implementation remains in the early stages. If recent (5-year average) federal funding levels continued and were matched by the state, construction of the remaining components of the congressionally authorized projects could take approximately 65 years; construction of projects in planning or those currently unplanned would further lengthen that timeline. At this pace of restoration, it is even more imperative that agencies anticipate and design for the Everglades of the future.

Incremental restoration progress from early CERP projects is difficult to evaluate because of a lack of rigorous assessment of outcomes relative to project goals and some limitations in existing monitoring plans. The committee reviewed available data and analysis on the restoration progress associated with three early CERP projects in which substantial project components are now in place and operating. The Picayune Strand Restoration Project shows increased water levels in the area of the two canals plugged to date. Hydrologic conditions are expected to improve further toward conditions at the reference sites once neighboring canals are plugged. Some early indicators of habitat response at Picayune Strand are apparent in the species composition of groundcover vegetation and suppression of some exotic species, but other ecological indicators, such as increased cypress regeneration, have not shown significant change. This lack of response could be due to lag times in ecological response, limitations in the monitoring plan, or insufficient hydrologic restoration to date. Analysis of these or other factors is an essential but missing component of performance assessment (see Chapter 4). At the C-111 Spreader Canal Project, neither hydrologic nor ecological response in Taylor Slough or Florida Bay due to the project has been documented based on monitoring data, because the monitoring and assessment plans are not robust enough to discern project impacts from existing hydrologic variability. The lack of specific numeric targets and an explicit plan and model to evaluate restoration progress hinders restoration assessment of these two projects. The Biscayne Bay Coastal Wetlands monitoring program has documented hydrologic and ecological responses, although both are limited by the small spatial scale of the components that have been implemented and important project components that are not yet constructed.

Concurrent project planning efforts have significantly advanced the CERP vision for water storage, but a holistic understanding of the benefits of the combined projects at a systemwide scale and their resilience to sea-level rise and climate change is lacking. Tentatively selected plans have been developed for the EAA Storage Reservoir and the Lake Okeechobee Watershed Restoration Project, which together propose adding 283,000 AF of surface storage and 80 ASR wells. Each project is expected to reduce high-volume discharges from Lake Okeechobee to the Northern Estuaries and to modestly improve the period that Lake Okeechobee stage is at ecologically preferred levels. The EAA Reservoir also provides moderate hydrologic improvements to WCA-2A and northern WCA-3A. By 2019, all of the large CERP storage projects at the northern end of the system will have been planned, with only the southern storage (i.e., Lake Belt) remaining unresolved. Preliminary modeling suggests that, with system optimization, the full storage planned in the original CERP may not be needed to provide the flows into the northern end of the Everglades as envisioned in the CERP. However, an integrated, systemwide modeling of the planned projects is needed to understand the combined benefits relative to restoration objectives. More rigorous analysis of the potential effects of climate change and sea-level rise on restoration outcomes is necessary in planning for all projects, so that restoration investments are designed for and more resilient to future conditions. The SFWMD and the Interagency Modeling Center have the talent and tools to conduct these analyses, and the SFWMD is pursuing this approach for planning and management issues outside of the CERP.

Impressive advances have been made toward water quality objectives in the stormwater treatment areas. The lowest flow-weighted mean total phosphorus concentrations to date (15 ppb for all STAs combined) were attained in water year 2017, and continued water quality treatment and science investments through the Restoration Strategies program are expected to further reduce phosphorus levels toward the 13 ppb goal. Achieving this goal is a necessary step to move forward with new water flows in the central Everglades. Understanding the dynamic ecological responses to restored flows (and the relative importance of phosphorus concentration and load in controlling ecosystem response) during these transitions is an emerging challenge. Where existing flows are currently being redistributed, as in the Decomp Physical Model and the non-CERP Florida Bay Initiative, project teams are following adaptive management approaches where feasible to learn from these efforts and to inform future Everglades flow restoration projects.

The recent completion of two major non-CERP projects is expected to provide important restoration benefits to Everglades National Park and increasing operational flexibility for managing high water events throughout the remnant Everglades. Completion of the Mod Waters and C-111 South Dade projects in August 2018 is a major achievement that has been more than 25 years in the making. Development of the Combined Operational Plan is under way, which will quantify the benefits provided by these projects.

4

Monitoring and Assessment

Monitoring is an essential aspect of ecological restoration efforts. Even though the need for monitoring is widely recognized, the National Academies of Sciences, Engineering, and Medicine (NASEM, 2017) noted that "most restoration projects in the U.S. and elsewhere often have lacked monitoring or the monitoring efforts have been insufficient to generate rigorous, decision-relevant, or publically accessible information." Large-scale analyses of restoration projects nationwide have found that many efforts lack measurable objectives and quantitative data suitable to evaluate restoration progress. Evaluations are also frequently hampered by inadequate monitoring designs, a disconnect between the collection of monitoring data and their synthesis to inform a subsequent decision-making process, funding challenges, and data management issues. Given the importance of monitoring to understand Everglades restoration progress, which is a key element of the committee's charge, this chapter examines Comprehensive Everglades Restoration Plan (CERP) monitoring at both the project and systemwide scales and offers recommendations to CERP agency staff as they work to improve those programs, ensuring that monitoring investments lead to useful information.

USEFUL MONITORING FOR RESTORATION

NASEM (2017) outlined three main purposes of restoration monitoring:

"(1) to assure projects are built or implemented and are initially functioning as designed (construction monitoring);

(2) to assess whether restoration goals and objectives have been or are being met (performance monitoring); and

(3) o inform restoration management, to improve design of future restoration efforts, and to increase ecosystem understanding (monitoring for adaptive management)."

These types of monitoring typically occur in sequence; initial construction monitoring gives way to performance monitoring, which in turn is necessary for adaptive management. These types may use some of the same metrics and even the same samples, but the monitoring plans are designed for different purposes. Although many projects also require monitoring for regulatory compliance (such as monitoring water quality for compliance with regulatory criteria), this chapter emphasizes performance monitoring and monitoring for adaptive management. These types of monitoring can be structured at either the project level or system-wide scale.

Performance monitoring provides the data needed to evaluate how well a project (or set of projects) is meeting its objectives, which may include specific hydrologic objectives as well as the expected ecological outcomes. Performance monitoring and subsequent evaluation ensures accountability to the funders of restoration, including taxpayers and funding agencies, by communicating the outcomes of restoration investments and providing input to decision making through adaptive management. NASEM (2017) judged that performance monitoring is "essential for all restoration projects." The CERP requires performance monitoring for all its projects.

Monitoring to support adaptive management is designed to address specific management questions or to fill restoration-related knowledge gaps, so that future Everglades restoration decision making or implementation can be improved by the knowledge gained (Nilsson et al., 2016). Adaptive management also fosters learning as new knowledge is gained on the ecosystem response to restoration efforts and the effect of changing conditions (e.g., climate change, sea-level rise) on restoration outcomes. Since 2011, all CERP projects have been required to develop a project-level adaptive management plan during the project planning process with specific monitoring requirements linked to decision-relevant uncertainties (USACE and SFWMD, 2011a). In 2015, RECOVER released a Program-level Adaptive Management Plan (RECOVER, 2015), but it did not include an associated monitoring plan or implementation framework.

Developing an effective restoration monitoring plan is challenging. The Monitoring and Assessment Plan (MAP) Assessment Strategy (RECOVER, 2006) outlines an approach to developing a monitoring program that is similar to that advocated by others (Convertino et al., 2013; Conyngham, 2010; NASEM, 2017; Pastorok et al., 1997; Thom and Wellman, 1996). The purpose of the MAP Assessment Strategy (RECOVER, 2006) is to "provide guidance to help ensure that the sampling designs and data analyses for the MAP monitoring components are adequate to detect measurable changes in hydrologic (including water supply and flood protection), water quality, and ecosystem indicators." The guidance provides a roadmap for the development of effective project-level monitoring.

Figure 4-1 outlines key elements of the development of a monitoring plan, as outlined in NASEM (2017) and consistent with RECOVER (2006).

Central to the existing guidance on monitoring (Conyngham, 2010; NASEM, 2017; NPS, 2012; RECOVER, 2006) is the idea that a strong monitoring plan will connect restoration goals, objectives, management questions, models, and the monitoring design. Restoration monitoring requires clear definition of restoration goals and numeric success criteria (measurable objectives) so that performance can be evaluated. A conceptual model of the system, perhaps supplemented by numerical models, is used to identify factors that are likely to yield useful information if they are monitored. Finally, determining the most important management questions that monitoring should address is necessary to define the role of adaptive management monitoring and the acceptable levels of uncertainty in judging project performance.

Once the monitoring purposes are established (i.e., in terms of informing decisions on construction, performance, and adaptive management), then numerous project considerations must be understood and factored into the restoration monitoring plan (Figure 4-1):

- *Specific indicators and metrics* should be identified that provide ecologically reliable information to address the fundamental questions and are often derived from the conceptual ecological models (e.g., Ogden et al., 2005). To be useful to policy makers and managers, indicators must be understandable, quantifiable, and readily interpreted (Box 4-1). The CERP staff already know much of this material (see RECOVER, 2006), but it is much easier to describe the best practice than to put it into practice.

- *Temporal and spatial scales of monitoring* also need to be considered (i.e., Where should samples be taken? How many? For how long? How frequently?). These decisions, which are part of the statistical sampling design, ultimately impact the rigor of conclusions that can be drawn from the data. A related element of the monitoring plan is the statistical model that will be used to evaluate hypotheses that are related to numerical criteria and hence objectives. Common statistical approaches for evaluation of restoration include design-based inference, generalized linear mixed models, multivariate methods, and Bayesian hierarchical models. Examples include estimation of a species' abundance, before-after control-impact (BACI) models, and nonmetric multidimensional scaling (see RECOVER, 2006). Power analysis and sample size calculations can be used to develop clear numerical objectives for the monitoring plan, prioritize metrics, and evaluate the likelihood of the plan's success. As described in NPS (2012), "sampling (or statistical) objectives specify information such as target levels of precision, power, acceptable Type I and II error rates, and magnitude of

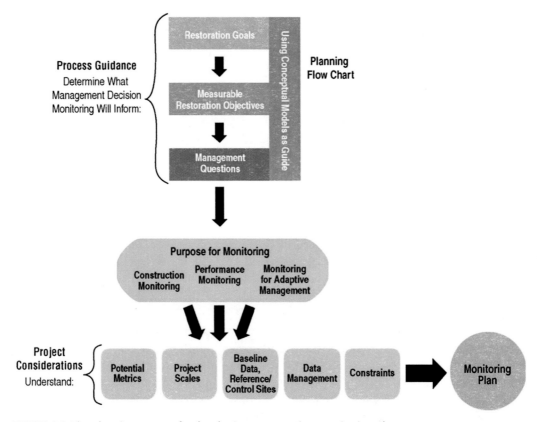

FIGURE 4-1 The planning process for developing a restoration monitoring plan.

SOURCE: NASEM, 2017.

---

change you are hoping to detect" (see Box 4-2). For example, a monitoring plan can be developed with a sampling regime that can detect an increase in roseate spoonbill nesting success of 20 percent with 80 percent certainty of detecting this change if it occurs and a 10 percent chance of false detection of an increase when it does not occur. Sampling can be adjusted if a higher level of certainty is required.

- *Baseline data and/or reference or control sites* will determine what responses can be attributed to restoration efforts amidst natural variability and ecosystem change. Assessment of existing baseline or reference data that may be appropriate for evaluating project objectives will inform plans to collect new baseline or reference data. Related projects may also benefit by jointly collecting

BOX 4-1
Principles for Developing Good Indicators

A set of principles developed from NRC (2000) can be used to evaluate currently used indicators to determine their effectiveness and their value added to our understanding of the system. Put simply, and all else being equal, few indicators are better than more indicators, and simple indicators are better than complicated indicators. These principles can also be used to manage the number of indicators within a monitoring plan. For example, could some be eliminated without losing much, or any, information? If so, then more resources may be available to reduce the uncertainty in the remaining indicators.

With respect to Everglades restoration, the most important principles are the following:

General importance. Does the indicator provide information about changes in ecological, hydrological, and biogeochemical processes that are judged to be important? At a systemwide scale, does the indicator integrate system response? Are some indicators redundant with others?

Conceptual basis. Is the indicator based on a well-understood and generally accepted conceptual model of the system to which it is applied? Does it respond predictably to system drivers? The conceptual model provides the rationale for the indicator, suggests what should be measured, and informs the interpretation of any changes that unfold.

Sensitivity. Is the indicator sensitive enough to detect important changes but not so sensitive that signals are masked by natural variability? Are the statistical properties well understood? This feature is especially important in ecosystems such as the Everglades that are notoriously variable in space and time. Understanding the magnitude and sources of variability in indicator measurement is critical in evaluating indicator sensitivity (Landres et al., 1999). In addition, changes in the variability of an indicator may be as important as changes in the mean value when assessing project performance.

Temporal and spatial scales. At what spatial and temporal scale does the indicator provide useful information, and how does that scale relate to the temporal and spatial scale of the project? In other words, an indicator might reflect a crucially important aspect of system structure or performance, but that aspect might change so slowly that monitoring it, even on an annual basis, might not be useful. Or, other aspects might be influenced by many variables over time and space and change so often that monitoring them might not be useful. As a result, the process of identifying and designing indicators involves understanding how often, and over what scale, they should be monitored and whether the scale is consistent with that of the project.

Data and resource requirements. How much and what kinds of information must be gathered to permit reliable estimates to be made of the indicator's values? Is gathering that information feasible given the financial and human resources available? In other words, even though the indicator might be useful, is monitoring it too expensive for the available budget? Many ecological indicators depend on data gathered over long periods, or over broad spatial scales for valid inference, so they require a long-term commitment to monitoring and to archiving the data.

Ease of communication. If an indicator is readily understood by the general public and therefore monitoring results can be clearly communicated to the general public, then the indicator will be useful in not only in decision making, but also obtaining support for decision making.

> **BOX 4-2**
> **Restoration Hypotheses and Statistics:**
> **An Example Using the Roseate Spoonbill**
>
> A statistically rigorous design will enable researchers to draw sound conclusions from the data with acceptable levels of confidence. Such a design incorporates decisions on testable hypotheses and plans for data analysis that affect the statistical power of the results. This hypothetical example describes the key elements of a statistical design for monitoring restoration success.
>
> *General Null hypothesis or H_0:* The restoration project does not lead to roseate spoonbill nesting success
> *General Alternative hypothesis or H_1:* The restoration project leads to roseate spoonbill nesting success
>
> *Metric*: Roseate spoonbill nesting success
> *Numerical criteria*: Nesting success is measured in terms of the number of chicks per nest that survive to a specific age. Success occurs if the mean increases to 1.5 chicks (or greater) per nest
>
> *Time frame*: Ten years
> *Spatial extent*: Two largest colonies sampled in each of five regions of Florida Bay
>
> *Specific null hypothesis H_0:* Mean nesting success is less than or equal to 1.5 chicks/nest
> *Specific alternative hypothesis H_1:* Mean nesting success is greater than 1.5 chicks/nest
> *Model*: A stratified or random sampling design with goal of estimating the mean chicks per nest over the region.
>
> *Possible statistical errors:*
> Type I—Conclude that the restoration is successful when it is not (falsely reject H_0). In this example, a Type I error would occur if it is decided that the average number of chicks per nest is greater than 1.5 when in fact it is less.
> Type II—Conclude that the restoration is not successful when in fact it is (fail to reject H_0). In this example, a Type II error would occur if it is decided that the average number of chicks per nest is less than or equal to 1.5 when in fact it is greater.
> *Power* is the probability of rejecting the null hypothesis when it is not true, and it is determined for a specific sample size, variability, and desired Type I error rate (usually 0.05). In this example, the power is the probability of correctly concluding that the average number of chicks per nest is greater than 1.5. For example, if the true mean number of chicks per nest is 2.0, the power would be calculated as the probability that the mean nesting success from sampled data is greater than 1.5 given the true mean is 2.0 for the spatial region and time frame.

baseline data or sharing a network of reference sites (NASEM, 2017), which will save monitoring costs.

- *Data-management procedures and protocols* should be established to ensure the accuracy and precision of the indicators. Data management procedures should be maintained throughout the restoration to standardize measurements and cross-calibrate instruments, especially when measurement technology changes. All methods should be documented so that people not associated with the original data collection can understand and reproduce the methods. All data used as input for indicators should be archived in as unprocessed, or raw, a form as possible so that the data will be available for computing a variety of indicators, many of which may be unanticipated. These procedures are especially important for long-term indicators. Data management should also address good practices such as quality assurance and quality control, careful construction of metadata, data publishing, and a means of data sharing (NASEM, 2017).

- Finally, *constraints*, such as long-term funding and accessibility of sampling sites, should be considered. Design decisions for the monitoring plan may depend on the availability of suitable reference sites or control sites. Other constraints related to social, political, or organizational factors must be considered (NASEM, 2017).

Because natural systems are complex over space and time, designing a truly useful monitoring program is much more difficult than is generally understood. For example, failure to recognize the difficulty of designing a good system of indicators typically results in a proliferation of indicators, many of which are of limited value, leading to a costly and unsustainable monitoring plan. Perhaps one of the greatest challenges in any monitoring program is identifying the aspects of the system's structure and performance that should be monitored.

CERP PROJECT-LEVEL MONITORING

Three CERP projects—C-111 Spreader Canal Western, Picayune Strand, and Biscayne Bay Coastal Wetlands—have made sufficient implementation progress, with some expectations for system response. In this section, the committee examines the ability of their monitoring programs to sufficiently document performance and identifies lessons learned for project-level monitoring across all projects. Within the CERP approach to development of a monitoring program, while generally solid, some differences in implementation may lead to mixed support for decision making about project success. This uncertainty is reducible.

Examination of Monitoring for Three Projects

A comparison of three CERP projects highlights some of the differences within and among the approaches to monitoring design. All three CERP projects involve performance monitoring, which is defined by the U.S. Army Corps of Engineers' (USACE, 2009) implementation guidance for monitoring ecosystem restoration as the "systematic collection and analysis of data that provides information useful for assessing project performance, determining whether ecological success has been achieved, or whether adaptive management may be needed to attain project benefits." A variety of publications describe the attributes of successful monitoring and assessment plans (Conyngham, 2010; Johnson, 2012; NASEM, 2017; Olsen and Robertson, 2003; Reynolds et al., 2016). Selected attributes for the three projects are summarized in Table 4-1, and the project-level monitoring plans are discussed in more detail below.

TABLE 4-1 Comparison of Key Features of Three CERP Project-Level Monitoring Plans

	Criteria	Biscayne Bay	C-111SC	Picayune Strand
Performance Monitoring	Establishment of conceptual ecosystem model	Yes	Yes	Yes
	Development of numerical project objectives: drivers (e.g., hydrology, fire regime)	Yes (hydrologic objectives)	No	Yes (hydroperiod)
	Development of numerical project objectives: ecological response	For some indicators	No[a]	No[b]
	Basis of assessment of project performance	Comparison with numerical objectives and trend assessment from baseline	Comparison with baseline	Comparison to reference
	Suitability of sampling design to determine statistically significant performance to date	Yes	No	No
Adaptive Management Monitoring	Includes evaluation of critical uncertainties and a process to assess monitoring results for adaptive management	Yes	No	No

[a] Numerical targets exist for some ecological indicators, but the contribution of the project to attainment of these indicator goals amidst several regional projects is not clear.
[b] The Picayune Strand Project uses reference conditions in neighboring Fakahatchee Strand as an objective, but numerical criteria for similarity to reference conditions that are viewed as success have not been defined.

Biscayne Bay Coastal Wetlands (Phase 1) Project

The Biscayne Bay Coastal Wetlands (Phase 1) monitoring and adaptive management plan is comprehensive, providing details on nearly all of the key elements shown in Table 4-1. Of the three projects reviewed here, this project is the only one with an adaptive management plan. The project encompasses three broad objectives: (1) restore the quantity, quality, timing, and distribution of freshwater to Biscayne Bay and Biscayne National Park, (2) improve salinity distribution and reestablish productive nursery habitat in the nearshore areas, and (3) preserve and restore natural coastal wetlands habitat (USACE and SFWMD, 2011b; see also Chapter 3). Monitoring is spatially stratified into three distinct ecological zones characteristic of Biscayne Bay: the near-shore bay, tidal saltwater wetlands, and freshwater wetlands. Within each of these zones, performance measures, monitoring parameters, and a range of possible adaptive management options are considered.

A conceptual ecological model was developed for the Biscayne Bay to provide insight into ecosystem functioning, advance the understanding of relationships between ecosystem components and external drivers and stressors, and show the hypothesized pathways of impacts (Browder et al., 2005). Some of these relationships have been investigated directly, for example, upstream water management and rainfall and freshwater flow into Biscayne Bay. Other hypotheses were developed in conjunction with the CERP conceptual model and its hypothesis clusters, for example to predict the response of submerged aquatic vegetation (SAV) to restoration of freshwater flows and establishment of mesohaline conditions.

The Biscayne Bay Coastal Wetlands (Phase 1) Project uses a mix of objectives. Targets for water delivery and quality (including salinity) are quantitative, while objectives for biota (e.g., shrimp, invertebrates, juvenile crocodiles) are qualitative or narrative (Table 4-2). Many of the biotic targets are linked to water flows and salinity stressor metrics: for example, one objective is for the nesting density for juvenile crocodiles in the bay to be similar to that in natural areas such as Crocodiles Lakes National Wildlife Refuge near Key Largo, with a corresponding salinity target of 0-20 psu (USACE and SFWMD, 2011b). However, no quantitative targets have been developed. Regarding SAV, there is an expectation that freshwater inflows to the bay will increase the cover of *Ruppia maritima* in the near-shore and coastal embayments and that the range of *Halodule wrightii* will increase along the shoreline (RECOVER, 2014), but, again, no quantitative targets for cover or distribution have been specified. In some cases, understanding of the system is not sufficient enough to make statements about the expected response: for example, the response of dwarf and transitional mangrove communities to increases in freshwater flows may be offset by sea-level

TABLE 4-2 Examples of Biscayne Bay Coastal Wetlands Objectives

Restoration Objectives	Indicators to Monitor Objectives	Time Period to Achieve Objectives
Hydrology: restore the quantity, timing, and distribution of freshwater to Biscayne Bay	Stage monitoring, hydroperiod, groundwater levels	Short
Salinity: 0 psu west of L-31E, 0-20 psu east of L-31, and a 10-25 psu near shore	Salinity	Short
Improve productive nursery habitat along the shoreline: e.g., oyster spat recruitment	Oysters (with salinity as habitat indicator)	Medium
Greater growth and abundance of submerged aquatic vegetation	SAV cover and species composition; salinity as habitat indicators	Medium
Increased cover and species composition of emergent wetland vegetation	Cover and species composition of macrophyte communities, wetland algae community composition	Medium
Increased crocodile nesting density similar to natural areas	Juvenile nesting density, salinity	Long

NOTE: Some of the indicators used to assess project success are shown, as is the relative amount of time necessary to achieve each objective. Short is 5-10 years, Medium is 10-20 years, and Long is >20 years.
SOURCE: USACE and SFWMD, 2011b.

rise. Coupling a model to quantify the hydrologic effects on salinity gradients could facilitate quantitative predictions that could then feed into the adaptive management framework. In the absence of explicit statements of the expected response, performance cannot be evaluated and rigorous testing of the project's efficacy is limited, thereby preventing implementation of the adaptive management plan. Although the project is in the early stages, it is not clear whether other monitoring questions are being actively investigated with monitoring data (e.g., how are juvenile crocodiles affected by water management?), nor is it clear whether the conceptual ecological model has been modified based on a refined understanding of the system.

A major objective of the Biscayne Bay Coastal Wetlands (Phase 1) Project is to improve productive nursery habitat along the bay's shoreline. One indicator of improvement is the reestablishment of relic oyster bars in the mouths of restored freshwater creeks. Oyster monitoring is planned for eight sites where significant increases in freshwater delivery to the bay are anticipated and includes measurement of oyster spat recruitment, oyster reef development, juvenile growth and mortality, and adult health and reproductive status. This effort requires sampling of an ambitious list of parameters, so the project team strategically identified which measures will receive priority for adaptive management actions (starting with salinity and oyster spat recruitment). The overall plan for oysters is robust, with the caveat that it cannot be fully implemented without setting quantitative targets for the indicators. This limitation may be linked to the lack of historical or baseline data upon which to build quantitative expectations (USACE and SFWMD, 2011b) and lack of data on the response of South Florida oyster populations to restoration efforts.

The monitoring plan includes some sample size analyses. For example, in the case of oysters, the sampling intensity for oyster density was selected to identify a shift in mean oyster density of 1.5 times the standard deviation with 95 percent confidence and nondetection rate of 20 percent (Type II error rate; see Box 4-2). As part of the adaptive monitoring plan for this metric, sample size can be adjusted based on first-year monitoring results. Although statistical models for analyzing the water quality data are not described in the monitoring plan, subsequent reports (e.g., Charkhian et al., 2018) describe the statistical methodology used for evaluating trends in water quality parameters.

For many of the Biscayne Bay Coastal Wetlands indicators, baseline (i.e., pre-construction) data are used to assess project-level performance (post-construction). For example, 2 years of baseline data have been used to compare the response of some aquatic species, such as fish, macroinvertebrates, shrimp, and crabs, to restoration efforts. This is a relatively short time period to gain an understanding of the annual variability in the system. In addition, water quality indicators, including salinity and temperature, have been monitored to establish baseline conditions and to quantify the relationship between freshwater flows and salinity patterns in the near-shore bay. Where possible, water quality sampling stations are located at the site of historical records to expand the time frame of baseline monitoring data. Vegetation monitoring operates at two scales: (1) detection of large-scale changes in vegetation based on aerial photo interpretation and (2) small-scale changes in vegetation by sampling plots located on transects along hydrologic gradients. To assess the large-scale changes in vegetation, a minimum of 20 years of aerial photos and mapping will be analyzed. Near the L-31E Project culverts, sawgrass was mapped in 2013 in a 370-acre area in the vicinity of culverts S-23A and S-23B to create a baseline map of sawgrass abundance. The decisions about baseline data collection were strategic, and, although variable, the temporal scope of the baseline data should enable assessment of whether restoration efforts are having the desired effect.

For each ecological zone in the adaptive management plan, a decision framework will link performance measures to monitoring, targets, and potential management options. However, despite the extensive effort devoted to developing the monitoring and adaptive management plan, reporting on restoration progress has largely been limited to hydrologic indicators (water levels and flows), water quality indicators (salinity, nutrients), and vegetation. Data on, or the assessment of, faunal indicators is lacking.

Multiagency analysis of project-level results that would support project-level adaptive management and public understanding of incremental benefits is also lacking. For example, the L-31E culverts have not consistently produced flows that meet regulatory targets, with particular deficits in dry seasons (see Figure 3-11) and limited biological responses (see Chapter 3), but a publicly available analysis of the causal factors has not been published. A delay in pump construction is likely a major factor; a temporary pump installed in October 2014 improved flows, although an interim pump operating since August 2017 has been less successful over the limited period for which data are available. Many factors may be affecting performance of this component to date, including interim pump capacity compared to the full project design capacity, availability of water, regional water management, and sea-level rise, which will necessitate greater head gradients to produce the intended flow. Because the interim pumps were installed recently, the data do not appear in the latest South Florida Environmental Report. However, analysis of the various factors affecting project performance since implementation, and of their likely impact on project goals, has been limited. Although the project is not complete, much could be learned from the existing monitoring to inform future actions, and these opportunities for learning are being missed.

Similar challenges are apparent at the Deering Estate component, which is complete and has been operating since 2012. Monitoring data for 2012-2016 showed that wetland water levels declined very quickly each day after nighttime pumping was stopped because of the highly permeable landscape. This large fluctuation in water levels was inconsistent with the original project design. Yet, nighttime pumping continued until 2016 when the SFWMD conducted an experiment comparing pulse versus continuous pumping to analyze pumping alternatives. Only in September 2018 did the SFWMD adjust operations to provide continuous pumping (M. Jacoby, SFWMD, personal communication, 2018). An analysis of the new lower but continuous pumping rates on the intended project outcomes for near-shore salinity has not been articulated.

Of the three projects reviewed in this section, the Biscayne Bay Coastal Wetlands most closely follows the guidance in RECOVER (2006), Conyngham (2010), and other monitoring documents cited above. In general, the monitoring plan addresses the necessary elements and serves as a model for the other CERP projects. However, the disconnect between the detailed plans for monitoring and adaptive management and the limited plans for analysis and reporting of project performance could limit what is being learned from the project and how it can inform future management actions. There is also a need to establish numerical targets for all project objectives and to provide a multiagency analysis of project performance relative to those objectives to ensure realization of the plan's full potential.

C-111 Spreader Canal Western Project

The monitoring plan for the C-111 Spreader Canal Western Project centers on three main project goals: (1) improve the quantity, timing, and distribution of water delivered to Florida Bay via Taylor Slough; (2) improve hydroperiods and hydropatterns in the Southern Glades and Model Lands; and (3) reduce ecologically damaging flows to Florida Bay and other receiving waters (USACE, 2016c). In a report evaluating project success, Kline et al. (2017) describe a broader set of objectives (Table 4-3). They note that long-term achievement of objectives will require additional freshwater input, which is partially controlled

TABLE 4-3 C-111 Spreader Canal Project Objectives, Related Indicators, and the Time Necessary to Achieve Each Objective

Restoration Objectives	Indicators to Monitor Objectives	Time Period to Achieve Objectives
Increased hydroperiod and freshwater conditions across the southern mangrove transition zone	Stage and canal stage Surface flow Groundwater level Salinity levels	Short
Increased coverage of brackish and freshwater submerged and emergent vegetation	Cover, biomass, and identification of vegetation	Short
Lower salinities in the southern lakes region	Salinity levels	Short
Increased abundance of the freshwater prey-based fish communities in the southern mangrove zone	Freshwater prey abundance and biomass	Medium
Increased productivity of the southern mangrove transition zone and northeastern Florida Bay	Salinity Submerged aquatic vegetation Prey fish abundance	Medium
Increased nesting success of roseate spoonbills in northeast Florida Bay	Nesting success	Long
Increased crocodiles	Growth and survival Relative abundance, and nesting success	Long

NOTE: Short is 5-10 years, Medium is 10-20 years, and Long is >20 years.
SOURCE: Adapted from Kline et al., 2017.

by other projects (Mod Waters, C-111 Spreader Canal [Phase 2], and the Central Everglades Planning Project).

To assess progress toward these three goals, the project will monitor salinity, vegetation, fish prey densities, stage, discharge, and water quality (Table 4-3). Historical pre-project data are available for hydrology, salinity, rainfall, fish abundance, and roseate spoonbill nesting metrics. National Science Foundation–sponsored Long-term Ecological Research (LTER) also provides baseline and continuing monitoring at five sites along a transect in Taylor Slough and at three sites in Florida Bay. Hydrologic, meteorological, and water quality indicators are monitored in and around the C-111 Spreader Canal Project area, and ecological and salinity monitoring is conducted in the southern coastal wetlands and Florida Bay (see Table 4-3).

Numerical restoration criteria are available for certain indicators[1] based on the ecological response to salinity in Florida Bay (Davis et al., 2005; Wingard and Lorenz, 2014), but it is not clear how these criteria quantitatively connect to the C-111 Spreader Canal Project. The current lack of quantitative objectives in the C-111 Spreader Canal project for flows in Taylor Slough or for salinity changes in Florida Bay makes it impossible to evaluate the project's restoration performance. Measurement of the project's effect on salinity in Florida Bay will require that project-related increased flows into Florida Bay are estimated and used to evaluate the reduction in salinity with and without the flow. The project's effect could be quantified in terms of projected reductions in violations of minimum flow targets that have been specified for the region affected by the project (NPS, 2012).

The performance of the C-111 Spreader Canal Project will be best evaluated using indicators that are clearly linked to the quantitative restoration objectives of the project and not to objectives for Florida Bay salinity more generally. Computer models, combined with the understanding of the ecosystem, could be used to develop quantitative objectives for flow in Taylor Slough and Florida Bay salinity at specific locations (perhaps near shore) based on the project benefits, and adjusted as necessary to account for the implementation of nearby projects that also provide benefits. Ecological models combined with restoration science knowledge could be used to develop quantitative ecological objectives in these regions. Monitoring data could then be compared against these objectives, using computer simulations and statistical models to help interpret the results amidst interannual hydrologic variability. Modeling and statistical analysis would likely reveal that project-related changes in salinity and SAV in areas of open water in Florida Bay will be difficult to detect through monitoring of the C-111 Spreader

---

[1] Mangrove community structure and spatial extent, water birds, prey-base fish and macroinvertebrates, crocodilians, and periphyton.

Canal Project alone. SAV and salinity are more appropriate indicators of a system-level response to the aggregate effects in the Florida Bay of all the CERP and non-CERP projects. Monitoring flows into upper Taylor Slough and eastern seepage back to the C-111 Canal and then comparing the resulting data to baseline data and quantitative objectives would provide more useful information about the project's performance. Although the project aims to reduce seepage and thereby increase flows in Taylor Slough to Florida Bay, no monitoring design document describes an approach to estimating the reduction in seepage or the increase in flows associated with the project and their uncertainty.

A quantitative analysis plan is needed to evaluate project performance, and long-term monitoring may be required to observe effects on ecological indicators in Florida Bay. It may be difficult to separate the effects of the C-111 Spreader Canal Project from those of non-CERP projects to the north that will also increase the deliveries of freshwater (e.g., ModWaters, C-111 South Dade, Florida Bay Project [see Chapter 3]). Coupling a hydrologic model that can quantify the effects of the C-111 Spreader Canal Project and other projects with a salinity model of Florida Bay could help to estimate the contribution of the individual projects toward the restoration targets. Although already in use in project planning, many of the necessary modeling tools are not being used with monitoring data to evaluate project performance.

In summary, the C-111 Spreader Canal Project, which began operations in 2011, has collected years of monitoring data. However, to date, monitoring and assessment are insufficient to understand the project's performance. Qui et al. (2018) summarized rainfall, stage, flow, and hydrology using graphical and descriptive methods, comparing the current year with pre-project years 2002-2012, but concluded that project success cannot be evaluated until all related projects in the area are completed and operational. However, in the interim, more could be done to assess the project's progress, such as developing numerical criteria, monitoring short-term indicators of project success, and implementing a quantitative analysis plan. Such improvements would also better support adaptive management efforts.

Picayune Strand Restoration Project

The monitoring plan for the Picayune Strand Restoration Project is comprehensive in its intent to determine whether the anticipated hydrologic, vegetative, wildlife, and estuarine benefits are being achieved. The project was developed to rehydrate a large drained wetland in southwestern Florida (see Chapter 3 for a description of the project and implementation to date). The monitoring plan was designed to address the nine project objectives listed in Table 4-4 (USACE and SFWMD, 2004a). There is no formal adaptive management plan, because

TABLE 4-4 Picayune Strand Objectives, Related Indicators, and the Time Necessary to Achieve Each Objective

Restoration Objectives	Indicators to Monitor Objectives	Time Period to Achieve Objectives
Reestablish natural flows to estuary	Not listed	Short
Increase surface aquifer recharge	Not listed	Short
Restore historic hydropatterns, sheet flow, and flow-ways	Water levels, hydroperiod	Short
Maintain sufficient water quality	Surface water quality, estuarine water quality, sediment, and tissue analysis	Short
Restore Everglades ecosystem to 1940s condition	Vegetation, nuisance and exotic vegetation, aquatic macroinvertebrates, fish, amphibians, oysters, SAV, oyster reef crab, nekton, wading birds, listed species (wood stork, panther, manatee)	Short/Medium to long
Restore ecological connectivity	Not listed	Short to Long
Provide resource-based recreational opportunities	Not listed	Short to Long
Restore natural fire regime	Not listed	Medium

NOTE: For purposes of this analysis, the time period to infer objectives was inferred from monitoring duration as presented in the Picayune Strand monitoring plan (USACE and SFWMD, 2009) and the judgment of the committee. Short is 5-10 years, Medium is 10-20 years, and Long is >20 years.

the project implementation report was developed before project-level adaptive management plans were required.

The project's objectives vary in whether they are articulated in qualitative or quantitative terms. For example, the target for oyster reef crabs is unusually detailed, while the target for vegetation is quantitatively vague (i.e., "comparable to the composition and structure of hydrologically similar reference sites"), with no indication of what degree of similarity between the restoration and reference sites is considered successful. Even when numerical criteria are established, as with hydroperiod and water levels by vegetation category (Table 4-5; USACE and SFWMD, 2009), they do not seem to be utilized to assess monitoring data (Barry et al., 2017; Chuirazzi et al., 2018; Worley et al., 2017).

The phased construction of the project, including the backfilling of several canals and construction of three pump stations (see Table 3-2 and Figure 3-4), led to development of a complicated monitoring schedule, with much of the post-construction ecological monitoring yet to occur. For the purpose of performance assessment, extensive hydrologic monitoring is available from 23 wells in Picayune Strand north of Tamiami Trail (since October 2003), three additional wells in brackish marshes south of Tamiami Trail (since November 2006), and

TABLE 4-5 Numerical Hydrologic Objectives for Plant Communities in Picayune Strand

PSRP Plant Communities	Hydroperiod (months)	Water Level (in) Wet	Water Level (in) Dry (1,10)
Mesic Flatwood, Mesic Hammock	≤1	≤2	-46, -76
Hydric Flatwood, Hydric Hammock	1 – 2	2 – 6	-30, -60
Wet Prairie, Dwarf Cypress	2 – 6	6 – 12	-24, -54
Marsh	6 – 10	12 – 24	-6, -46
Cypress	6 – 8	12 – 18	-16, -46
Swamp Forest	8 – 10	18 – 24	-6, -36
Open Water	>10	≥24	<24, -6
Tidal Marsh, Mangrove, Beach	Tidal	Tidal	Tidal

SOURCE: USACE and SFWMD, 2009.

24 wells along two transects across Fakahatchee Strand (since 1987). Vegetation and aquatic macroinvertebrate monitoring was completed in 2016 following the plugging of the Merritt Canal.

The existing analyses of hydrologic data involve comparison to reference site hydrology or inference of trends and are only qualitative in nature (see Figures 3-6 and 3-7). Quantitative comparisons of observed hydroperiods to the target conditions for different vegetation types would provide a clear and concise early indicator of the degree of hydrologic responses and whether the appropriate pre-conditions for vegetative community shifts had occurred. If specific hydrologic metrics are not considered in performance assessment and only comparisons to reference are made, then a statistical pattern-matching technique should be applied to a comparison of restored sites to reference sites with the desired vegetative communities. Trend methods (Skalski et al., 2001) that evaluate parallelism between reference and restored sites have been used effectively to evaluate restoration success; recovery is considered complete when the restored site begins to track or parallel the reference site. Bayesian methods (Conner et al., 2015; Prato, 2005) may also be useful for evaluation of restoration.

For the biotic components of vegetation and aquatic macroinvertebrates, the Picayune Strand monitoring plan uses comparison to reference sites to demonstrate restoration success, with an ultimate target of matching the distribution, composition, and extent of 1940s vegetation. Baseline vegetation sampling occurred from 1996 to 2005 at approximately 90 transects located in reference areas (in Fakahatchee Strand Preserve State Park and Florida Panther National

Wildlife Refuge) and across the entire project area. Subsets of transects in the project area are sampled to assess the impacts of various phases of the restoration. For example, vegetation was sampled in 2005, 2008, 2009, 2011, and 2012 at transects to the east of the Merritt Canal to assess the effects of various stages of the Prairie Canal restoration, and the 2016-2017 sampling was the first event for Merritt Phase transects post-restoration (see Figure 3-5). Baseline sampling for treefrogs, aquatic macroinvertebrates, and fish occurred in 2005-2007. In 2016 a subset of these transects were sampled to assess changes in vegetation and aquatic fauna (16 project transects [8 pairs] that were classified as having either full or partial hydrologic restoration and 11 reference transects). The 8 pairs of restored transects represented six vegetation habitat types: cypress, wet prairie, pine flatwoods/hydric, pine flatwoods/mesic, hammock/hydric, and freshwater marsh. The 11 reference transects were chosen to represent three types of pre-drainage habitat: cypress, wet prairie, and pine flatwoods/hydric.[2]

A basic two-sample t-test is used to compare a range of vegetation parameters in reference and restored sites in different vegetation habitats (Barry et al., 2017). While valuable, two issues with this approach should be addressed to reduce uncertainty in evaluating project success. First, the approach to infer similarity with reference sites is based on an individual sampling event and does not account for the natural variability in reference conditions over time and space. This variability in reference sites could be evaluated and used to determine when the restoration at restored sites can be considered successful. Second, statistical inference for Picayune Stand is hampered by small sample sizes for both the restoration and reference sites. The number of sample sites within a habitat group is currently small (1-4) in the areas with full or partial hydrologic restoration (Barry et al., 2017), which makes it difficult to reject the hypothesis of no difference (because of a lack of power in the testing process). However, as additional project components are implemented and a larger area becomes affected, more transects should become available for sampling. Larger samples sizes can help to identify small but significant differences, as can longer monitoring periods. In general, the same issues of natural variability in reference sites and small sample size apply to sampling of frogs, fish, and macroinvertebrates. For example, an overall increase in Cuban treefrog populations (an invasive species) at reference sites makes any distinction between restored and reference sites difficult to ascertain, and the relatively short time period under restored conditions makes changes difficult to detect (Worley et al., 2017).

While the Picayune Stand sampling scheme is generally robust in its spatial and temporal extent, temporal expectations for restoration performance are gener-

---

[2] The reference transects included three transects of cypress only, plus one transect of cypress with graminoid understory, four transects of wet prairie, and three Pine flatwoods/hydric transects.

ally left unstated. The lack of specific temporal expectations combined with the lack of numerical objectives make it difficult to effectively evaluate restoration success, which, in turn, limits the capacity for adaptive management (coupled with the lack of a formal adaptive management plan). For example, while groundcover wetland affinity indices in restored sites show a response in cypress and prairie communities, but not in pineland communities. Is this to be expected? How should vegetative expectations be adjusted in light of hydrologic responses to date? Significant learning has already occurred through hydrologic monitoring, including the recognition of the zone of influence of unplugged canals on hydrologic response. However, numerical targets for hydroperiod or more specific hydropattern matching analysis between restored and reference locations, combined with stated expectations for vegetation response under restored hydrologic conditions, would better support a formal process of evaluating why/why not expectations for ecological response have been met. Currently, significant learning opportunities are being missed.

In summary, the Picayune Strand Restoration Project approach to monitoring is comprehensive for a phased restoration effort, but the lack of numerical targets, rigorous statistical approaches, and articulation of temporal expectations and envelopes of variability limits the ability to assess performance and take advantage of learning opportunities.

Key Lessons Learned from Project-Level Monitoring

The comparison of the monitoring programs for three early CERP restoration projects reveals several findings. A strength of Everglades monitoring has been the collaboration and consistency in goals and methods. This consistency reduces uncertainty in how data are collected by various agencies and should reduce uncertainty in information used in management decisions. The comparison of monitoring also reaffirms several lessons that are consistent with RECOVER (2006) that could improve future project-level monitoring programs, including

- Use of quantitative objectives and envelopes of expectations,
- Careful selection of indicators and appropriate scales of project monitoring,
- Rigorous statistical and computational analysis, and
- Missed opportunities for learning through adaptive management plans.

Quantitative Objectives and Envelopes of Expectations

Clear, quantitative objectives or expectations of system response are essential components of project-level performance monitoring. In the three projects examined, quantitative objectives are clear (Biscayne Bay Coastal Wetlands), varied and sometimes unclear (Picayune Strand), and lacking (C-111 Spreader Canal).

Performance targets may be specific (e.g., water depths at or above 1 foot for 30 consecutive days) or indicate an envelope or range of acceptable performance (water depths between 1 and 2 feet for 60 percent of the growing season). Defining restoration targets as a range of values under which an ecosystem may vary, instead of as a fixed target, results from an increasingly sophisticated understanding of how ecological processes vary over time and space. Early in the development of the science of ecological restoration, ecosystems were thought to be deterministic, returning to pre-disturbance conditions when stressors were removed, and therefore success could be measured against static targets. More recently, the awareness that ecosystems show thresholds and nonequilibrium dynamics has led to the definition of envelopes of anticipated response (Falk et al., 2006). For example, prior to the disruption of Everglades hydrology, salinity levels in Florida Bay would have fluctuated according to the natural variation in annual rainfall patterns and temperatures. Goals that acknowledge this variability in the natural system reflect the reality of ecosystem dynamics and allow learning to take place, resulting in a more effective approach to restoration. When an indicator falls outside of the expected behavior range, the project team should evaluate why the outcome is not responding according to the conceptual model used to select the indicator.

The Biscayne Bay Coastal Wetlands Project's approach considers variability and therefore defines an acceptable envelope of monitoring results. In this case, salinity monitoring is used to determine whether the appropriate salinity envelope is being achieved for eastern oysters at or near the mouths of major tidal creeks. The target envelope has been set at 10-25 psu, which ensures adequate flows during the dry season to maintain salinity at or below 25 psu. Because eastern oysters cannot tolerate low salinity beyond short periods, the target is designed to limit freshwater flows so that salinity does not fall below 5 psu for more than 5 consecutive days (USACE and SFWMD, 2011b). By using a range of expectation for restoration end points, the "precision problem" in restoration monitoring can be avoided (Hiers et al., 2016).

In Picayune Strand, assessment of project performance would be strengthened by explicitly stated, quantitative expectations of the restoration response, including the time expected for restored sites to reach reference conditions. For example, this project seeks to reestablish or expand three major plant community types (pine flatwoods, wet prairie, and cypress), which are expected to have differential responses over time. Currently, the expectation is only that the trajectory of ecosystem response will trend toward reference conditions, with no time frames for response. Quantitative objectives for hydroperiod (or a numerical target for hydropattern matching) would enable statistically rigorous early performance assessment. In the C-111 Spreader Canal Project, some quantitative indicators for the broad restoration efforts in the region exist, but project-specific measurable objectives are lacking. Clear quantitative project objectives, with estimated timescales for achieving them, are essential to the success of project-level adaptive management.

Monitoring programs may not have established quantitative objectives at the project planning stage, and substantial uncertainty about quantitative objectives may persist through project design. This uncertainty should be clearly described, and as the project evolves and knowledge is gained about ecosystem response, the quantitative objectives can be refined. For example, as the Kissimmee River Restoration Project evolved, it moved from early, narrative objectives ("restore ecological integrity to the Kissimmee River and its floodplain") to quantitative, measurable indicators to track progress toward that goal ("percent of water years that the mean depth at broadleaf marsh sites is greater than 1 ft for 210 days consecutively in a water year") (Koebel et al., 2017). It is only with specific targets that progress can be measured and goals operationalized.

Careful Selection of Indicators and Appropriate Scales of Monitoring

The ability to assess a project's success depends on the appropriate selection of indicators and their associated metrics. Considerations include measurement uncertainty, timescales of response, natural variability, and the influence of non-restoration-related factors (such as disease). For example, vegetation cover will respond to increased water from a project but also to rainfall magnitude, duration, frequency, and timing. Episodic events, such as hurricanes and exotic species invasions, may also have a strong influence. Monitoring plans must consider the likelihood that a metric can detect success given potential confounding factors.

Identifying appropriate indicators to quantify objectives is challenging when system changes occur over large areas and long time periods. The scale of the monitoring plan should align with the scale of the anticipated effects of that project and not be conflated with measures of systemwide or regional response. For the C-111 Spreader Canal Project, existing indicators in Florida Bay and the southern mangrove transition zone may be unable to detect project-related changes in the face of natural variability and may be better indicators of long-term regional restoration improvements. Indicators of flow and seepage near the project features are needed to assess project performance and inform adaptive management.

In some cases, indicators must be able to detect slow change that occurs over the long term. For both Biscayne Bay Coastal Wetlands and Picayune Strand, vegetation with a long life cycle, such as cypress, pine flatwood communities, and mangroves, are being monitored, which may require 10 or more years before solid indications of restoration progress emerge. For example, tree basal area, which has a long response time, is being measured in the five Biscayne Bay long-term vegetation monitoring plots. If these slowly changing indicators are judged to be important to performance assessment, larger samples sizes can help identify small but significant differences. Otherwise, longer monitoring periods may be necessary to see a significant effect, and less frequent monitoring of these indicators will reduce costs. Thus, the selection of indicators should recognize the spatial and temporal complexity of the responses they are intended to measure.

The costs of monitoring and the complexity of data management and interpretation increase exponentially as the number of indicators increases. The parsimonious selection of indicators can increase the effectiveness of a monitoring program; that is, fewer indicators can be better. The key question is how to reduce the number of indicators without losing useful information. Recognizing that not all aspects of the system can be tracked, the Biscayne Bay Coastal Wetlands Project used the conceptual ecological model to select a species (e.g., oysters) to monitor whose abundance is directly linked to system stressors (Browder et al., 2005). In this way oysters act as a surrogate for other parameters. The monitoring schemes of the three projects reviewed here are ambitious with many, sometimes overlapping indicators. This can lead to the problem of being data rich and information poor; that is, project managers have a surplus of data that is difficult to translate into useful and actionable information (Doren et al., 2009).

Rigorous Statistical and Computational Analysis

The project-level monitoring programs lack rigorous computational analysis, either in the design of the monitoring plan or in the analysis of the data, or both. For example, the Picayune Strand and the C-111 Spreader Canal monitoring plans do not address how collected data will be analyzed. As a result, it is unclear whether the monitoring data can lead to meaningful conclusions about project performance. Future projects should return to monitoring fundamentals that were described in RECOVER (2006) and USACE planning (Conyngham, 2010) and include initial determination of the statistical or other type of model(s) that will be used to evaluate project performance. Then, the modeling approach should be evaluated using simulated or historical data to determine whether the monitoring and sampling, especially the spatial and temporal extent of sampling and number of samples, will result in sufficient data to render conclusions about project performance with known certainty.

Kline et al. (2017), for example, developed an analysis process to assess the performance of the C-111 Spreader Canal Western Project after the monitoring was completed. Although the results of the analyses are not incorrect, such a process might be viewed as subjective because the decisions about the analysis were not part of the monitoring plan (i.e., see Conyngham, 2010). Similarly, Robinson et al. (2016) used rainfall pattern matching to identify a year from a pre-project historical dataset that is comparable to the current year. They then compared the data for the current and selected historical year for hydrology, submerged aquatic vegetation, and fish metrics. They made decisions about the spatial extent of the rainfall (which gauges to use), how rainfall data were summarized (monthly mean), and the time period for evaluation (water year), as well as the methodology for determining which year in the prior monitoring period was most similar to a current year. The approach involves choice of a measure of similarity, method for summarization (multidimensional scaling), and number of dimensions to use. Again, this approach may be appropriate, but different researchers might use a different approach a posteriori and come to different conclusions. For example, Kline et al. (2017) used the same approach but could not find an adequate match because of the unique rainfall pattern in water year 2016. Defining the approach to statistical analysis in the monitoring plan reduces uncertainties and allows for an a priori estimate of the strength of the inference that is made about project success.

Evaluation of individual project performance with multiple, neighboring projects may require the use of more complex statistical models or hydrologic and ecological simulation models as part of the data analysis. Even with these tools, some ecological indicators may be difficult to evaluate given natural variability in the system and will require a long-term commitment to monitoring. In addition, it may be impossible to separate project-level effects from the effects of other neighboring projects, and therefore the performance of multiple projects may need to be judged collectively. Upfront statistical analysis could identify these metrics and scale issues, so that planners are aware of the long-term investments needed to assess project performance.

Factors such as climate variability, saltwater intrusion, and long-term changes in rainfall frequency, intensity, timing, and duration are likely to influence the assessment of project performance. It may be unrealistic to assess hydrologic and ecological objectives under different climate scenarios without a strong modeling approach that enables evaluation of ecological responses given multiple stressors and changing baselines. Simple statistical modeling may not be able to provide strong inference, making more-complex modeling necessary, for example, by combining statistical and ecosystem simulation. Further development of predictive models may be needed to evaluate restoration success under scenarios of change.

Missed Opportunities for Learning

Without adaptive management plans, the opportunities for learning about the restoration process are limited. Of the three CERP projects reviewed here, only Biscayne Bay Coastal Wetlands has an established adaptive management plan; Picayune Strand was initiated before adaptive management plans were required, and the C-111 Spreader Canal Project team has not developed one. The absence of an adaptive management plan makes it difficult to structure monitoring and evaluation so that new knowledge can be applied in a flexible decision-making process. Performance monitoring may show that project objectives are not being met, but it cannot provide the reasons for failure or suggest corrective actions, nor can it resolve uncertainties in the understanding of the system (NASEM, 2017; RECOVER, 2015). The opportunity for learning through monitoring is also limited by the lack of integration of modeling with monitoring, which can aid in setting quantitative objectives and predicting reference conditions.

Learning is also restricted by the fragmented nature of reporting for specific projects and the absence of multiagency assessment of project-level monitoring, as RECOVER performs for systemwide monitoring. The assessment of hydrologic conditions is reported in at least three separate reports for Picayune Strand that use different analysis methods.

SYSTEMWIDE MONITORING

The CERP Monitoring and Assessment Plan (MAP) was originally developed in 2001 and has been developed and revised several times, with the latest version in 2009 (RECOVER, 2009). It was intended to establish a "robust and scientifically defensible monitoring program." According to RECOVER (2009):

> "[i]nitially, the MAP had three broad objectives: (1) establish a pre-CERP reference condition (e.g., the condition prior to implementation of restoration activities and significant anthropogenic changes associated with CERP) including variability for each of the performance measures; (2) provide an assessment of the system-wide responses of CERP implementation; and (3) detect unexpected responses of the ecosystem to changes in stressors resulting from CERP activities."

The systemwide monitoring plan also provides data that are used communicate the state of the system to the public. RECOVER uses the System Status Reports (SSRs; RECOVER, 2010, 2014) to provide information related to these three broad objectives, and the South Florida Ecosystem Task Force uses the data to show the status of its systemwide indicators (Box 4-3). Although the monitoring plan is designed for these systemwide objectives, some projects use the data to supplement project-level performance monitoring. All three projects reviewed in this chapter use some RECOVER systemwide data to evaluate project performance.

RECOVER (2009) and earlier documents provide details about the plan's development. In brief, RECOVER started with conceptual ecological models and hypothesis clusters to identify uncertainties and frameworks for understanding how aspects of the ecosystem might respond to environmental changes and restoration activities. MAP 2009 reemphasized the importance of being able to detect change in status and trends and to understand *why* those changes are

---

BOX 4-3
Systemwide Ecological Indicators Developed by the Task Force

The Task Force Science Coordination Group developed systemwide indicators to assess and communicate the success of restoration at a systemwide scale. Considerations used to assess the suitability of indicators included whether there was a conceptual ecological model, how well the indicators mapped to ecological and physiographic features in the Everglades, and the degree to which they integrated ecological processes. As a result of these processes, the Science Coordination Group chose 11 ecological indicators plus 3 indicators relating to water quality, hydrology, and flood control:

- Invasive Exotic Plants
- Lake Okeechobee Nearshore Zone Submerged Aquatic Vegetation
- Eastern Oysters
- Crocodilians (American Alligators & Crocodiles)
- Fish & Macroinvertebrates
- Periphyton
- Wading Birds (White Ibis & Wood Stork)
- Southern Coastal Systems Phytoplankton Blooms
- Florida Bay Submerged Aquatic Vegetation
- Juvenile Pink Shrimp
- Wading Birds (Roseate Spoonbill)

The Task Force reports biennially on the status of these indicators using stoplight indicators (red, yellow, and green based on the status and trends).

SOURCE: Brandt et al., 2014.

occurring. MAP 2009 stated that "addressing the *why* questions are essential" to successful implementation of adaptive management.

Several publications describe the desirable features of systemwide monitoring, including two by the National Research Council (NRC, 2000, 2003a). The development of the MAP was a large effort in which consideration was given to identify the characteristics of effective indicators and metrics (NRC, 2003a). From the beginning, CERP partners and prior National Academies committees (NRC, 2003a, 2007, 2010, 2012) have recognized the importance of comprehensive monitoring and assessment to the success of Everglades restoration. Beginning with a dedicated workshop in November 2001, National Academies committees have reviewed the development of MAP and the selection of appropriate and practical performance measures by RECOVER (NRC, 2003a, 2007, 2008, 2010). NRC (2008) concluded that "[t]he number of performance measures is not inherently problematic" but noted that "the set of performance measures should be reviewed regularly to determine whether . . . adequate data collection for each could be sustained over the course of the restoration." NRC (2010, 2012) called for a comprehensive review process to update the systemwide monitoring program based on new information and lessons learned from the monitoring program to date; changing ecosystem conditions, including climate change; and changing budget conditions for RECOVER and other agencies that supply systemwide monitoring in support of the CERP.

RECOVER's 5-Year Plan includes an update of the Systemwide Monitoring Plan. Although a full review of the current MAP was not feasible during this study cycle, the committee received several briefings on the systemwide monitoring program (see Figure 4-2) and discussed the challenges to developing a cost-effective systemwide monitoring program with staff from other large ecosystem restoration programs. Key lessons are offered here to inform RECOVER's upcoming review.

Value of Adaptive Monitoring

For the Everglades, a large systemwide monitoring data set now exists that is available to inform the future path of restoration, but to derive the most value from monitoring investments, it is important to ask whether the systemwide monitoring plan addresses the most critical program needs going forward. Several environmental restoration projects have moved toward monitoring approaches that emphasize adaptation as new knowledge emerges about how the ecosystem operates and how new challenges impact the restoration effort—known as adaptive monitoring (Lindenmayer and Likens, 2009). Adaptive monitoring approaches can incorporate new technologies (e.g., drones, sensors, or satellite

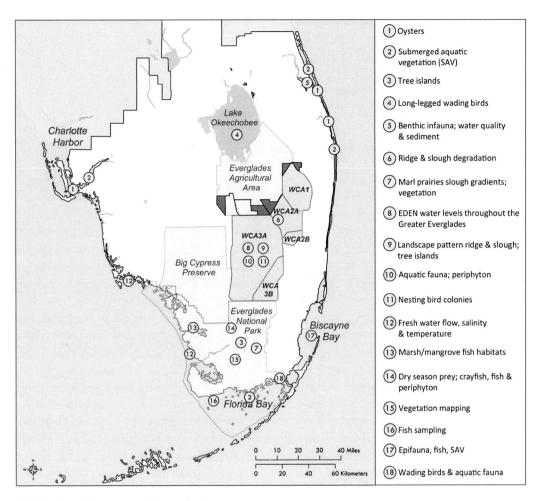

FIGURE 4-2 CERP systemwide monitoring plan.

SOURCE: Adapted from A. Patterson, USACE, personal communication, 2017.

imaging technology) and analysis strategies to improve efficiencies, but more importantly, to maintain the relevancy of the monitoring program to management. Periodically revisiting a monitoring program provides the ability to incorporate emerging stressors that may severely impact the probability of restoration success, such as changing precipitation patterns, regional invasive species, or sea-level rise, which may not have been apparent when the monitoring program was initially developed. Adaptive monitoring strategies can also be used to update the statistical designs of monitoring plans to provide information that meets the desired certainty for decision makers. The value of the data produced to date and the expected value of future data to address program needs moving forward are periodically reviewed to ensure that they continue to provide maximum value, in terms of information, to the decision-making process.

Process Considerations for a Monitoring Review

Given the considerable investment in monitoring that typically occurs at the beginning of a restoration effort, it can be difficult to accommodate change and flexibility as the restoration program matures. Fortunately, several programs have tried different approaches to adapting their monitoring programs to evolving needs and can offer lessons learned as the CERP looks to the near future (2019) when a review of its systemwide monitoring programs is scheduled to begin.

Because of current funding realities in some programs, which include delayed budget processes and frequent last-minute appropriations, a ranked priorities map for monitoring elements based on collective scientific and decision-making input has proven to be extremely valuable. For example, a comprehensive review of Chesapeake Bay's monitoring was initiated in 2009, which was the first in more than 20 years. The review revealed a significant disconnect between the program objectives and the monitoring investments. Although monitoring of Bay conditions had intensified over time, critical information on the effectiveness of management actions in the watershed had not been compiled. Because the monitoring budget was kept stable, monitoring had to be realigned between Bay and watershed efforts to respond to the decision-making needs. To do this, a specific ranking of priorities for funding/de-funding had to be agreed upon; the strength of the process used to do so is reflected in the fact that the funding/de-funding map has remained in use for 9 years and is only now being revisited (Chesapeake Bay Program, 2009).

The committee sought input from three environmental restoration monitoring programs (Chesapeake Bay, San Francisco-Bay-Delta, and Albemarle-Pamlico Sound) about their experiences with adaptive monitoring. Their processes included three important common elements:

1. A formalized commitment to periodically review the effectiveness of the monitoring program and make adjustments,
2. The presence of independent experts in some capacity, and
3. Inclusion of both key stakeholders (e.g. decision makers, managers, modelers) and monitoring experts at the table.

The programs addressed these elements in various ways. The commitment to periodic reviews ranged from formal presence in regulatory orders to internally motivated requests. The use of independent experts ranged from external monitoring review by an independent science advisory board to provide broad guidance for monitoring revisions to inclusion of outside experts in the monitoring review process. The stakeholders involved also varied by program. Because modeling is a critical tool in the Chesapeake Bay restoration and monitoring supports model refinement, modelers were important participants in the process. In addition, because states are responsible for restoration implementation, state resource managers were critical partners. The inclusion of stakeholders in the monitoring review also assisted identification of potential monitoring partners; the Chesapeake Bay review identified more than 200 monitoring efforts that could be integrated into the program to eliminate redundant efforts, resulting in significant cost reductions.

Ultimately, the conversation that must take place is between decision makers and monitoring experts, and it may be iterative in nature. Stakeholders will need to identify the monitoring information that will best address their immediate decision-making needs and articulate their risk tolerance. Monitoring experts will need to respond with a range of options (both data types and statistical designs) for meeting those needs, along with the uncertainty that each poses. Stakeholders will then have to determine the appropriate balance between the cost and value of monitoring information given the available options.

Overall, representatives from each of these large restoration programs judged the process of reviewing the monitoring programs to be of significant value. The benefits included ensuring that monitoring can address emerging uncertainties, eliminating redundancies, and offering an informed funding map for future monitoring. Commitment to such a process avoids the unintended drift in large monitoring programs that results in a growing deficiency of information for relevant decision making.

CONCLUSIONS AND RECOMMENDATIONS

Monitoring is essential to assess the effectiveness of ecosystem restoration efforts (i.e., what was the response?) and support adaptive management (i.e., if the expected outcomes did not occur, why not?). The collection and assessment of monitoring data are necessary to communicate the outcomes of restoration efforts to decision makers and the public, support learning from the restoration outcomes, and guide decisions about future changes that may be needed. The committee's conclusions and recommendations for monitoring were informed by a review of project-level monitoring for three early CERP projects (Picayune Strand, Biscayne Bay Coastal Wetlands [Phase 1], and C-111 Spreader Canal Western) and of the CERP systemwide monitoring program. Although this and previous National Academies committees have recommended improvements in CERP-associated monitoring programs, this does not necessarily mean that additional funding for monitoring is required. There are many ways to improve both the efficiency and the effectiveness of the CERP monitoring program within the existing monitoring budget.

The three CERP projects analyzed vary in the extent to which they have implemented effective monitoring plans. The RECOVER 2006 Assessment Strategy for the Monitoring and Assessment Plan provides valuable guidance on how to establish monitoring plans to detect change and evaluate progress toward goals. However, the three projects have not implemented this guidance systematically. For example, there is variation in whether quantitative restoration objectives are articulated. Not all projects have established a clear sampling design and data analysis plan as part of the monitoring plan, which could limit the usefulness of the results.

Quantitative restoration objectives, with accompanying expectations of how and when they will be achieved by management actions, should be developed for each project during the project development process. Quantitative objectives are needed to effectively measure restoration progress and operationalize goals. In addition, an acceptable level of variability of monitoring data around these objectives should be established so that management actions can be adjusted and adapted if the desired outcome is not being achieved. In the early stages of project development (i.e., before the monitoring plan is designed), project teams may be more comfortable with narrative objectives. However, it is essential to establish quantitative objectives as part of the monitoring plan with uncertainty described as appropriate. As programs evolve, more is learned about project functioning, and as knowledge and modeling tools improve, the quantitative objectives can be refined.

Monitoring plans should include an evaluation of the ability to detect restoration success given natural variability and sampling constraints. Models and historic monitoring data can be used to select metrics and design sampling plans to determine restoration success with a high degree of certainty, considering natural variability, expected changes from factors such as sea-level rise, and constraints such as site accessibility, funding, and personnel. These analyses should help to direct monitoring investments to where they will be most effective.

Modeling and statistical tools should be used in combination with monitoring data to assess restoration performance. External factors, such as precipitation and temperature variability, impact hydrologic and ecological responses, making it difficult to determine ecosystem response to restoration projects when compared to baseline data. Where feasible, reference and control sites can be used to quantify project-related effects, but for most Everglades projects, well-characterized reference and control sites are not available. Additional tools, such as modeling and statistical analyses, are essential to help quantify the effects of the projects and to separate them from ongoing system variability and trends. Modeling tools can be used to separate the effects of other long-term changes, such as sea-level rise or invasive species, on project performance as well as to understand the effects of an individual project within a region that is affected by multiple, interacting projects. Although the CERP has a strong modeling program for project planning, models are rarely used to interpret monitoring data, greatly reducing the potential value of existing data. When numerical or statistical models are to be used in performance assessment, the data analysis plan should be identified before the data are collected to reduce bias in the assessment.

Project-level monitoring should be revisited periodically to ensure that sampling designs and data-analysis plans are effective and efficient and that monitoring investments yield useful information. Periodic reviews would include assessing the usefulness of the monitoring data to meet decision-making needs and the relevance of the selected indicators to the questions being asked. Other considerations include the validity of the conceptual model, the timing and rate of ecosystem response relative to sampling intervals, the adequacy of the spatial scale of monitoring considering the scale of anticipated response, and the use of rigorous computational or statistical tools for data analysis. Such reevaluation should lead to more effective and efficient performance monitoring and will strengthen the capacity to learn through adaptive management.

The full implementation of adaptive management plans will substantially increase learning about the restoration process. Adaptive management allows learning to take place as new knowledge is gained about ecosystem response to restoration and how changing future conditions (e.g., climate change, sea-level rise) might affect restoration outcomes. Only one of the three CERP projects analyzed (Biscayne Bay Coastal Wetlands) has an established adaptive management plan. Without an adaptive management plan, it is difficult to structure monitoring and evaluation so that new knowledge can be applied in a flexible decision-making process. Performance monitoring may show that project objectives are not being met, but performance monitoring alone cannot explain the reasons for failure or inform restoration decisions. Learning through monitoring is also limited by the lack of integration of modeling with monitoring, which can aid in setting quantitative objectives and projecting reference conditions. Monitoring plans for adaptive management should evaluate whether the restoration project is expected to result in measurable change with high certainty for adaptive management indicators and over what time frame.

The CERP program currently lacks a mechanism for multiagency assessment and reporting of project-level monitoring results. The RECOVER System Status Reports (SSRs) provide comprehensive multiagency analysis and synthesis of systemwide monitoring and assessment of trends, but they do not provide analysis and assessments of individual project performance. Currently, most reporting of project-level monitoring data occurs via the South Florida Environmental Reports (SFERs), which annually compile the data associated with permit monitoring. However, these reports contain limited analysis of long-term trends, project performance relative to expected objectives, and potential adaptive management needs. Additionally, the SFERs do not provide the opportunity for multiagency perspectives or RECOVER input. A variety of other reports, many by contractors, also provide sometimes fragmented summaries of data from monitoring but information on overall project performance relative to objectives remains lacking. A better-organized, multiagency analysis and assessment of project performance based on monitoring results should be developed to provide transparency to decision makers, funders, and stakeholders. This effort will also help support project-level adaptive management efforts.

The upcoming RECOVER review of its systemwide monitoring plan should be embraced as an opportunity to improve its effectiveness and efficiency. Many of the same issues addressed in project-level monitoring, such as the ability of the sampling plan to address the key questions and the availability of data to allow adaptation of management actions if the desired outcomes are not being achieved, are evident in current approaches to systemwide monitoring. The monitoring review, scheduled to begin in 2019, should also consider the relevance and usefulness of indicators, statistical rigor of the assessment, use of modeling for data analysis, and the appropriateness of the spatial and temporal sampling design to ensure that the investments in monitoring are being made toward data that can inform assessments and decision making. Scientists should understand and incorporate the needs of decision makers into the monitoring program. Similarly, decision makers should understand what information scientists can and cannot provide. This will require an iterative two-way dialogue between managers and scientists covering such issues as risk tolerance or aversion, what amount of confidence in data summaries is acceptable and possible, which indicators are most important and feasible to monitor, what decisions the information will be used for, and what information is of most scientific value for specific decisions. The process by which monitoring reviews are performed requires a thoughtful and intentional approach, such as the inclusion of stakeholders, modelers, and independent monitoring experts in the review process. Periodic systemwide reviews of monitoring should be incorporated into the work plan of RECOVER so that the monitoring program remains effective and appropriate in the years ahead.

5

Lake Okeechobee Regulation

Lake Okeechobee is a shallow impounded lake at the center of the greater Everglades that provides a myriad of services critical to South Florida (Figure 1-2). It is the largest freshwater lake in the southeastern United States, with a surface area of 730 square miles (mi^2) and a volume in excess of 4 million acre-feet,[1] and it is the largest component of water storage in the South Florida ecosystem (NRC, 2005). The water storage provided by the lake, the quality of that water, and the manner in which it is released to downstream ecosystems, all have implications for the success of the Central Everglades Restoration Plan (CERP). The amount of water stored in Lake Okeechobee during the wet season affects the frequency and magnitude of regulatory discharges to the Northern Estuaries. The high water events and algal blooms of 2016 and 2018 provided recent reminders of the central role of water storage and water quality in Lake Okeechobee on ecological conditions in the Northern Estuaries. Storage in Lake Okeechobee also strongly determines the availability of water in the dry season for Everglades restoration and urban and agricultural water supply. Additionally, storage has implications for habitats and resident biota within the lake, including wading birds, a commercially valuable sport fishery, and the endangered Everglade snail kite.

In the next few years, decisions will be made about the provision of storage by Lake Okeechobee as part of an upcoming Lake Okeechobee Regulation Schedule review, scheduled for 2019-2023 (USACE, 2018b). The lake regulation schedule was lowered in 2007 primarily to reduce the risk of catastrophic failure of the Herbert Hoover Dike while rehabilitation efforts were implemented. As a result, hundreds of thousands of acre-feet of potential storage capacity were lost, with adverse implications to the Northern Estuaries, dry season Everglades flows, and water supply (NASEM, 2016) but positive implications to lake ecology (USACE, 2007b). The upcoming Lake Okeechobee Regulation Schedule revi-

---

[1] See https://www.sfwmd.gov/our-work/lake-okeechobee.

sion will provide an opportunity to evaluate the feasibility and the benefits and risks of allowing higher water levels in the lake once the Herbert Hoover Dike repairs are complete to recapture some of that storage. Recent reports (Graham et al., 2015; NASEM, 2016) have emphasized the importance of water storage to meet the original Everglades restoration goals and to adapt to possible future changes in precipitation.

All of the major attributes of the regional system are considered when new lake regulation schedules are evaluated, and the end result inevitably involves tradeoffs. Analyses of tradeoffs require that the costs, risks, and benefits of the alternatives are well understood, both within the lake as well as to water supply, the estuaries, and the remnant Everglades ecosystem. This chapter seeks to provide insights into the first portion of the tradeoff equation, by examining the impacts of storing an additional 0.5-1.0 foot of water in Lake Okeechobee on the ecology of the lake. The South Florida Water Management District (SFWMD) specifically requested that the committee evaluate the effects of higher water levels on the biota of Lake Okeechobee to help inform future deliberations on water management. The committee recognizes that changes in lake management must be done in a regional system context, because the lake is only one part of a complex, interconnected, water-management project and regional ecosystem.

In this chapter, the committee summarizes the major factors that affect the lake ecosystem, based on the best available data, and highlights additional monitoring, modeling, and research that could be used to reduce uncertainties related to lake management decisions. The chapter begins with an overview of Lake Okeechobee water quality and an update of the Herbert Hoover Dike rehabilitation project, which have important implications for a new regulation schedule. Next, the committee provides an overview of processes by which lake levels affect lake ecology, including uncertainties and key research needs. The committee recommends an ecological monitoring approach that could help to support operational optimization under any regulation schedule, as well as the use of recently developed modeling tools to support a regulation schedule review. The chapter concludes with a brief discussion about tradeoffs in the regional water system, including lake ecology, storage, the Northern Estuaries, Water Conservation Areas (WCAs), the Everglades, water supply, navigation, and flood management.

WATER QUALITY

The water quality of Lake Okeechobee affects the ecological condition of both the lake and downstream ecosystems. Release of large amounts of water from the lake can contribute nutrients and seed the formation of toxic algae blooms in the Northern Estuaries (Phlips et al., 2012). The movement of phosphorus-rich water from the lake to the south necessitates more extensive treatment to meet water quality standards before it can be used for Everglades restoration. Phosphorus pollution in Lake Okeechobee is a tremendous challenge in respect to substantive remediation of the South Florida ecosystem (Dunne et al., 2011).

Phosphorus enrichment of the lake dates back to the mid-1900s. Historically, cattle ranching was the main agricultural use of the watershed north of the lake, but from the 1950s to the 1960s, dairy farming increased eight-fold, with a corresponding increase in phosphorus exports from 250 to 2,000 metric tons/year (Flaig and Havens, 1995). When combined with improved drainage throughout the watershed, nutrients were quickly transported from sources to wetlands, rivers, and ultimately Lake Okeechobee. Many of the historic sources have been remediated, and loads at point sources have declined (Julian et al., 2013, 2014, 2015, 2016; Payne and Xue, 2012; Piccone, 2010, 2011). However, a large amount of legacy phosphorus remains throughout the watershed (Dunne et al., 2011), which must be immobilized or allowed to purge from the watershed over time for loads to the lake to decline.

The most recent 5-year (2013-2017) average annual load of total phosphorus to Lake Okeechobee (531 metric tons/year) greatly exceeds the Total Maximum Daily Load (TMDL) for total phosphorus of 140 metric tons/year. The TMDL was established to achieve a lake water concentration of total phosphorus below 40 parts per billion (ppb) and improve the structure and functioning of the ecosystem (FDEP, 2001; Havens and Walker, 2002). There is considerable year-to-year variation in total phosphorus loading to Lake Okeechobee largely because of differences in water inflows associated with meteorological conditions (Figure 5-1). Only in extreme drought years, when little water enters the lake, have loads approached the TMDL.

Total phosphorus loading to the lake has not significantly declined over the 1974-2017 period of record (Figure 5-1), despite a large array of projects that have reduced phosphorous sources (Flaig and Havens, 1995; Julian et al., 2013, 2014, 2015, 2016; Payne and Xue, 2012; Piccone, 2010, 2011). The lack of response of loads at the watershed scale reflects the accumulation of legacy phosphorus in the watershed. This legacy phosphorus, which is slowly migrating downstream through soils, sediments, and wetlands, has a mass of approximately 160,000 metric tons (Dunne et al., 2011). It is estimated that if the remaining contemporary phosphorus sources (e.g., active dairies, cattle ranches, vegetable farms) were eliminated, the legacy phosphorus alone is sufficient to maintain a loading rate near 500 metric tons/year for as long as 50 years (Dunne et al., 2011).

Concentrations of total phosphorus in the open water zone of Lake Okeechobee increased from the period of earliest measurements in the 1970s,

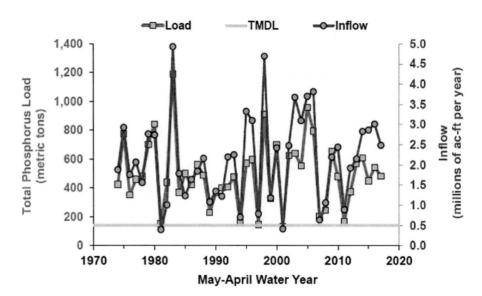

FIGURE 5-1 Time series total phosphorus loads and inflows to Lake Okeechobee compared to the Total Maximum Daily Load (140 metric tons/year).

SOURCE: Zhang and Welch, 2018.

reaching a peak of 233 ppb in 2005 (Figure 5-2) following three major hurricane impacts on the lake (Havens et al., 2011). After those hurricane impacts had subsided in 2017 (Ji et al., 2018), the in-lake concentration of total phosphorus was 150 ppb, still far in excess of the goal of 40 ppb. The long-term increases in total phosphorus concentrations in the lake likely are a combined result of the high concentrations in inflowing waters, mobilization of phosphorus from the sediments, and a decreased ability of lake sediments to remove phosphorus from that water (Havens et al., 2007).

Mass balance calculations for total phosphorus conducted since the 1970s show that Lake Okeechobee has been a net sink—more phosphorus enters the lake than leaves it—as phosphorus sorbs to the lake sediment. Net sediment retention of total phosphorus was very high in the early years of monitoring but has slowed considerably since the mid-1990s. The lake continues to display an overall net retention of total phosphorus, but at a lower rate.

Sediment phosphorus retention is a function of inflow, outflow, and water residence time, and it is affected by phosphorus binding in lake sediments. When phosphorus-poor sediments are buried under more recent sediments with

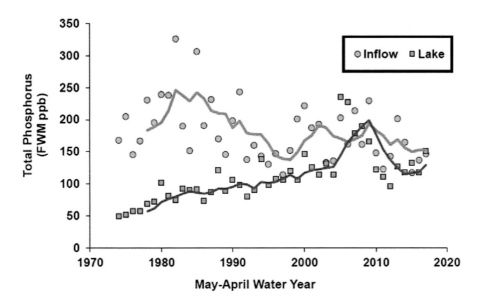

FIGURE 5-2 Time series total phosphorus concentrations entering and in the water column of Lake Okeechobee. The solid lines are 5-year running average concentrations.

SOURCE: Zhang and Welch, 2018.

---

a higher phosphorus concentration, those surface sediments have less capacity to sorb phosphorus from the water column. It is common for lakes that have experienced a long history of high phosphorus loading to shift from being a sink to a source of phosphorus (Sas, 1989; Sondergaard et al., 1992). At some point after external loads are greatly reduced and phosphorus-rich sediments are buried with more recent "clean" sediments, this interaction can reverse and the sediments can become a sink for phosphorus. Even with that change, it is estimated that after reaching the phosphorus loads in the TMDL, it could take as long as 40 years for total phosphorus in the water column to appreciably decline (James and Pollman, 2011). As of 2018, a decline in total phosphorus has not even begun, and it will not start until years after the external loads to the lake are reduced.

HERBERT HOOVER DIKE REHABILITATION

In 1930, Congress authorized the Herbert Hoover Dike, which now encircles most of Lake Okeechobee with 143 miles of embankment, five inlets/outlets, nine navigation locks, and nine pump stations. The capacity of water to flow into the lake greatly exceeds the capacity to flow out, and after large rain events, runoff can result in a rapid increase in lake level. Water levels in the lake are regulated by the U.S. Army Corps of Engineers (USACE) based on a regulation schedule that is a set of seasonally varying rules guiding lake operations. If lake level exceeds an upper boundary set by the regulation schedule, water must be released to reduce the risk of failure of the Herbert Hoover Dike (Figure 5-3).

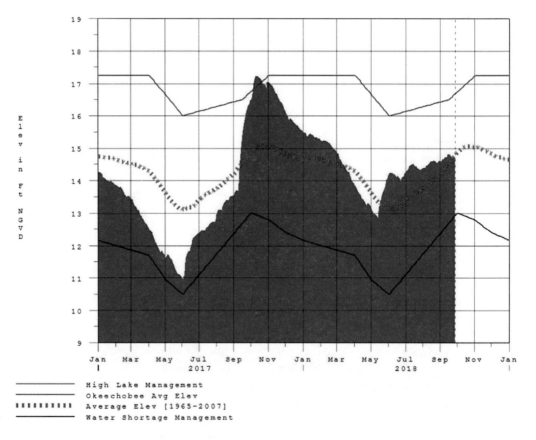

FIGURE 5-3 Hydrograph of surface elevations in Lake Okeechobee, showing the rapid rise in water level after Hurricane Irma resulted in heavy rainfall on the lake and over the watershed to the north of the lake. Within a 6-month period, the lake went from approaching water shortage management to exceeding high lake level management thresholds.

SOURCE: http://w3.saj.usace.army.mil/h2o/plots/okehp.pdf.

Lake level is allowed to rise prior to the winter dry season to ensure that the amount of water is adequate for downstream agricultural irrigation and for urban uses in South Florida. Lake level is lowered before the summer wet season to provide for maximal storage capacity to accommodate heavy rain events and tropical storms that may occur. Extreme rain events, however, can quickly push the lake level above what is considered safe at a particular time of the year (Figure 5-3). Failure of the embankment would cause massive damage and loss of life. In 2004, the USACE classified the Herbert Hoover Dike as Level 1 (i.e., highest risk) with regard to safety, and a major rehabilitation project has been under way since 2007.

The Herbert Hoover Dike rehabilitation project included a 21.4-mile cutoff wall in the embankment in an area called Zone A (Figure 5-4) that was completed in 2013. An ongoing project to construct an additional 6.6 miles will complete the cutoff wall in Zone A by 2022. In addition, culverts are being replaced around the perimeter of the lake. A USACE dam safety modification report (USACE, 2016d) identified the final measures needed to reduce intolerable risks in the remaining reaches of the Herbert Hoover Dike, based on the current lake regulation schedule. Work in Zones B through D were identified based on the relative probability of failure associated with internal erosion or overtopping of the embankment. Rehabilitation measures proposed include a 24.5-mile cutoff wall in Zone B, a 4.1-mile cutoff wall in Zone C, armoring of bridge abutments in Zone C, and embankment flood walls in Zone D (Figure 5-4). To date, the USACE has invested over $1 billion in the rehabilitation work, which is estimated to cost more than $1.8 billion in total (USACE, 2018g).[2] Completion of the project was anticipated by 2025, but in July 2018, the USACE announced $514 million in supplemental funding to expedite the Herbert Hoover Dike rehabilitation (USACE, 2018b). With USACE supplemental funding from the Disaster Relief Requirements Act of 2018, intended to reduce risk from future floods and hurricanes (Public Law 115-123), combined with projected fiscal year (FY)2019 funds and $100 million in state funds, the USACE estimates that the Herbert Hoover Dike rehabilitation project will be completed in 2022.

While the rehabilitation project is under way, Lake Okeechobee is being operated under a protective regulation schedule (Lake Okeechobee Regulation Schedule [LORS] 2008). As with prior schedules, LORS 2008 proactively pushes water out of the lake in advance of the hurricane season, but this schedule lowered the seasonally variable bands that determine when and how much water is to be released (Figure 5-5). On average, the LORS 2008 keeps the lake 1 foot lower than the earlier Water Supply/Environmental (WSE) regulation schedule.

---

[2] See http://www.saj.usace.army.mil/Missions/Civil-Works/Lake-Okeechobee/Herbert-Hoover-Dike/.

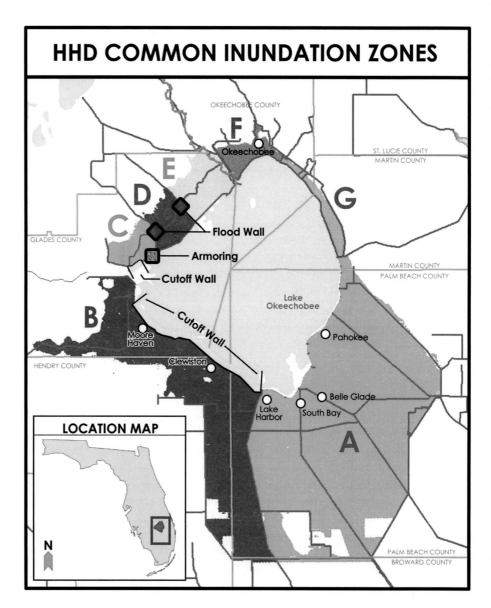

FIGURE 5-4 Locations where additional cutoff walls, armoring, and flood walls will be constructed in the Herbert Hoover Dike Rehabilitation project, with a current estimated completion date of 2022. Letters indicate common inundation zones designated surrounding the lake. The rehabilitation project has already installed 21.4 miles of cutoff wall in Zone A.

SOURCE: USACE, 2016d.

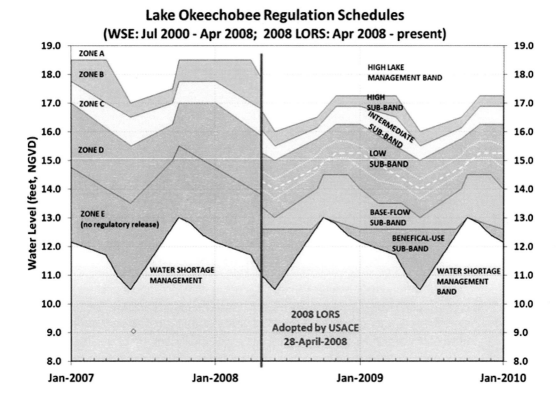

FIGURE 5-5 A comparison of the current Lake Okeechobee regulation schedule in comparison to the previous schedule, showing the lowering of flood release zones that results in approximately 460,000 to 800,000 acre-feet less potential water storage, depending on the time of year.

SOURCE: SFWMD.

The recent management schedule has the potential to remove between 460,000 and 800,000 acre-feet of storage from the regional system at any given time compared to the lake regulation schedule under which the CERP was developed (NASEM, 2016). USACE (2016d) noted that any revisions to LORS 2008 would require an updated Herbert Hoover Dike risk assessment, which could be undertaken concurrently with rehabilitation efforts. Because the storage capacity of Lake Okeechobee affects the amount of storage needed in other parts of the regional system, NASEM (2016) recommended that the process to revise LORS 2008, then scheduled for completion in 2025, begin as soon as possible. The July 2018 Draft Integrated Delivery Schedule (USACE, 2018a) advances this timeline, with a scheduled start date in 2019 and completion in 2023, 1 year after the scheduled Herbert Hoover Dike completion.

LAKE HYDROLOGY AND ECOLOGY

The upcoming review of the Lake Okeechobee Regulation Schedule will consider many factors including dam safety, environmental benefits and impacts to the Northern Estuaries and the remnant Everglades, effects on water supply, and ecological impacts within the lake. This section provides an overview of the key characteristics of Lake Okeechobee that influence the effects of changing lake levels on lake ecology and discusses potential ecological effects of a deeper lake.

Conditions That Modulate the Effects of Altered Lake Levels

To fully discern how changes in the magnitude, timing, and duration of high and low water affect the ecosystem, it is critical to understand some unique properties of Lake Okeechobee that influence water quality, currents, shearing stress of wind and waves, and horizontal transport of nutrients. Those conditions fall into three general categories addressed below: morphometry of the lake bottom; impoundment of the lake; and lake sediment type, hydrodynamics, and spatial distribution of nutrients. Throughout these discussions, *lake level* (or stage) refers to the surface elevation of the lake above mean sea level, while *depth* refers to water depth at a particular location.

Morphometry of the Lake Bottom

When viewed from a satellite image, the lake appears to have just two zones—a western littoral zone with emergent vegetation and a large pelagic zone with a homogeneous open expanse of water. However, five distinct zones occur in the lake, largely controlled by the bottom topography, water column depth, sediment type, and proximity to major inflows (Figure 5-6; Phlips et al., 1993a). The *central pelagic zone* has a relatively flat bottom and a water column depth that typically is 14 to 16 feet, depending on the surface elevation of the lake. Along the south and west shoreline is a shelf where the water depth is shallower—in the 3- to 5-foot range. In this *near-shore zone*, tens of thousands of hectares of submerged aquatic vegetation (SAV) can occur, including both vascular plants and benthic macro-algae (Havens et al., 2002; Hwang et al., 1999). The near-shore zone is separated from the central pelagic zone by an ancient limestone reef that sometimes is exposed under low lake conditions. Between

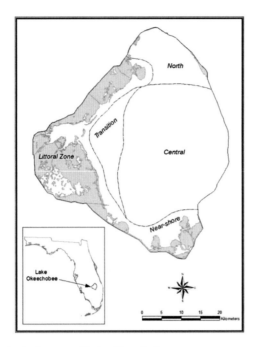

FIGURE 5-6 Five ecological zones of Lake Okeechobee.

SOURCE: Phlips et al., 1993a.

---

much of the near-shore zone and the central pelagic zone is a *transition zone* where underwater irradiance and nutrient conditions are optimal for formation of algal blooms. Westward of the near-shore zone is the *littoral zone*, with water column depths of 1 to 3 feet and emergent, floating, and SAV. At times, depending on inflow conditions, the *north pelagic zone* can be quite distinct from the central pelagic, resulting in a fifth zone with high nutrient concentrations and dissolved organic color from riverine inflows.

Historically Lake Okeechobee had no natural impoundment except for some limited areas of organic berm created when the lake level was very high because detritus pushed up onto the land. When periods of heavy rain caused the lake to rise, the lake spread outward to the west and south into the Everglades (McVoy et al., 2011), becoming a deeper but also much larger lake with an expansive littoral zone extending far to the west and south. There always were shallow water areas for fish, wading birds, and other biota that required shallow flooded habitat.

Impoundment of the Lake

When Lake Okeechobee was encircled by the Herbert Hoover Dike, its natural configuration was lost and the lake became a large shallow basin with a steep rim. In the impounded lake, a littoral zone developed in the shallowest area along the western shore. This zone accounts for approximately 35 percent of the lake surface area and has a diverse assemblage of plants, with sawgrass, spikerush, and surface periphyton mats, much like the Everglades. When the lake level rises to 15 feet in the impounded lake, water meets the edge of the dike, and the entire littoral zone has standing water. When the lake rises higher, the littoral zone is flooded with deeper water. Depending on antecedent conditions and the duration of flooding, this can have adverse effects on vegetation, fish, and other biota that are described in detail later in this chapter.

Lake Sediment Type, Hydrodynamics, and Spatial Distribution of Nutrients

Historically most of the area encompassed by the Herbert Hoover Dike had sand sediments, with peat sediments at the southern end of the lake. Because of agricultural activity and the straightening of the Kissimmee River upstream, the central pelagic zone has accumulated nearly 5 million cubic meters of organic mud sediments (Fisher et al., 2001) that are rich in phosphorus and easily entrained into the overlying water column by wind and waves (Havens et al., 2007; Jin and Sun, 2007). As a result, the central pelagic zone often is highly turbid, with low light penetration, light-limited primary production (Aldridge et al., 1995), and total phosphorus concentrations in excess of 100 ppb (Havens et al., 2007).

The near-shore zone, overlying sand and peat, may have considerably greater light penetration, nutrient-limited phytoplankton, total phosphorus in the 30 to 50 ppb range, and widespread SAV (Havens, 2003). Total phosphorus concentrations in the interior region of the littoral zone can fall below 10 ppb.

The spatial heterogeneity in phosphorus concentrations diminishes as lake level rises from 15 to 17 feet or more. When lake level is below 15 feet, circulation is constrained to the central pelagic zone by the southern and western limestone reef, and heterogeneous conditions can occur between the pelagic and near-shore zones. As the lake rises from 15 to 16 feet, the circulation of nutrient and sediment-rich water from the pelagic zone extends across the entire pelagic zone including the near-shore zone, and the lake becomes homogenized, with similar concentrations of total phosphorus and suspended solids, and low light penetration (Havens et al., 2007). When the lake level is at 17 feet, the nutrient- and sediment-rich water is transported into the littoral zone (Aumen and Wetzel, 1995; Jin and Sun, 2007).

Potential Ecological Effects of a Deeper Lake

A revised lake regulation schedule potentially could allow for deeper water in the lake. Predicting how the ecosystem would respond to deeper water is complex, because effects of water depth on critical habitat and biota are related to time of year, duration of the increased depth, rate of change in depth, and antecedent conditions of the biota (robust, recovering, or impacted). There has been considerable research, involving both controlled experiments and long-term assessment, to document the effects of hydrology on the ecology of Lake Okeechobee (Havens, 2002; Havens and Gawlik, 2005; Johnson et al., 2007), and yet uncertainties remain because of the complex influencing factors. The following sections summarize the best available information for each of the major ecological zones, with an identification of uncertainties and research needs, where applicable.

Pelagic Zone

The pelagic zone, including both the central and northern portions, is the largest and deepest part of the lake. This region generally is highly turbid, cannot support SAV, and its water quality is largely determined by wind-driven resuspension of phosphorus-rich sediments. The central pelagic zone occasionally has widespread and intense blue-green algae blooms (Havens et al., 2016), but these are rare with occurrence related to prior high inputs of nutrients during a wet period followed by hot calm conditions. Maintaining a high lake level does not appear to adversely affect the ecology of the pelagic zone.

Near-Shore Zone

The near-shore zone has two different states—a clear-water state with abundant SAV and a turbid, phytoplankton-dominated state without SAV (Aumen and Wetzel, 1995), similar to other shallow eutrophic lakes in the temperate zone (Scheffer, 1989; Scheffer et al., 2001). When nutrients enter the near-shore zone while it has abundant SAV, those plants and their associated periphyton sequester phosphorus at a rapid rate (Hwang et al., 1999), which appears to suppress phytoplankton from forming blooms. When SAV is sparse or absent, nutrient inputs can stimulate blooms of phytoplankton, particularly cyanobacteria, in the near-shore zone (Phlips et al., 1993b). There is a delicate balance between a clear state and a turbid state, depending on factors including high water levels. A catastrophic event such as a major hurricane can shift the near-shore zone of Lake Okeechobee from the clear to turbid state (Havens et al., 2011). Once that happens, recovery to a clear state may not occur without some counteracting event, such as a drought (Ji et al., 2018).

The near-shore zone of Lake Okeechobee provides a number of ecosystem services. It supports a recreational fishery, is an area for bird watching, and, when it has a high density of plants, is an area that sequesters nutrients from the lake water and prevents near-shore algal blooms (Aumen and Wetzel, 1995). Water level, the seasonality of water level changes, and the rates of change could affect a myriad of plants and animals in this zone (Havens and Gawlik, 2005; Johnson et al., 2007). Research has shown that by assessing the effects on two components of this zone—SAV and bulrush—effects on other biota can be inferred.

Submerged aquatic vegetation. SAV provides critical ecosystem services (Aumen and Wetzel, 1995), stabilizing lake sediments, preventing resuspension, and reducing water column turbidity. The near-shore zone with its SAV is a transitional location for a variety of fish species, as they start life in the littoral zone and subsequently move into the near-shore and then the pelagic zone as they mature (Fry et al., 1999). These fish, in turn, provide a food source for wading birds.

Major factors controlling the spatial extent and growth of SAV in the near-shore zone are sediment type and light availability. The latter is affected both by depth and light-attenuating particles in the water column. Research has quantified the light requirements of different species of SAV in Lake Okeechobee (e.g., Grimshaw et al., 2002), the recovery of SAV from high water stress (Havens et al., 2004; Steinman et al., 2002), and the capacity of the seed bank of the near-shore sediment to provide resilience to SAV after periods of loss (Harwell and Havens, 2003).

Submerged aquatic vegetation modeling efforts in Lake Okeechobee have included both simple (e.g., Harwell and Sharfstein, 2009; Havens et al., 2002) and complex (e.g., Jin and Ji, 2013) approaches. The SFWMD recently documented a strong inverse relationship between the spatial extent of SAV and the minimal lake level between May and August (the growing season) (Figure 5-7). In years when lake level decreased to 12 feet, the spatial extent of total SAV was three times greater than years when the lake depth decreased to only 15 feet. These results suggest that achieving that yearly low water level is important to having widespread SAV in Lake Okeechobee. Although the relationship between spatial extent of total SAV and minimal annual depth was highly significant, for the vascular plant component of the SAV (which is most important as habitat for fish), depth explained just 45 percent of the year-to-year variation. It is not clear what other factors accounted for the remaining 55 percent. Further, little is known about how the SAV will respond to a change in depth at a particular time, because of the aforementioned influence of turbidity, wind energy, and

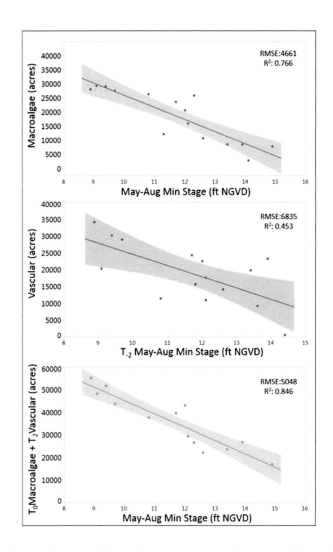

FIGURE 5-7 Regression models based on yearly vegetation maps of Lake Okeechobee, from 1999 to 2016. The shaded bands represent the 95 percent confidence interval on the regression. The spatial extent of macro-algae is inversely related to lake level. The spatial extent of vascular plants is inversely related to minimal lake level 2 years prior (reflecting the longer growth/response time of those plants). Taken together, the minimal lake level can explain greater than 84 percent of the variability in annual submerged plant coverage in the lake.

SOURCE: Zach Welch, SFWMD, personal communication, 2018.

antecedent conditions on that response. For example, in a year when the SAV is robust following prior years of drought, the effects of not attaining a low lake level of 12 feet might be minimal compared to a year when the SAV is depleted by prior years of high water. Further research is needed to discern with greater certainty how the SAV in Lake Okeechobee responds to changes in water depth and lake level.

Near-shore emergent plants. Emergent plants including bulrush (*Schoenoplectus* spp.) occur in bands along the western shoreline of Lake Okeechobee, providing ecosystem services such as serving as a substrate for invertebrates consumed by various species of fish, producing seeds valuable as food for many species of water birds, attenuating wave energy that protects the edge of the littoral zone, and serving as a refuge where SAV can grow on the leeward side of the emergent plants (Aumen and Wetzel, 1995; Coops et al., 1996). In Lake Okeechobee, this emergent plant community is rooted at about 10 feet in elevation, so stems of 6 feet in length or less are fully submerged at a lake level of 16 feet. In many emergent vegetation species such as *Schoenoplectus* spp., their stems consist of very porous plant tissue that allows for gas transport from the atmosphere to the anoxic soils, where it supplies oxygen to the roots and can detoxify potentially harmful compounds such as hydrogen sulfide. By definition, emergent plants are rooted in the soil, typically with the lower portions of their stems under water and the upper portion of their stems, leaves, and reproductive organs above water. If the plants are fully submerged, the exchange of oxygen and carbon dioxide gasses between the shoots and roots is prevented, eventually leading to plant death. When fully submerged, the stems would be exposed to greater wave energy, which could result in mechanical damage to the plants (Cronk and Fennessy, 2001; Vymazal, 2011). Although the committee was unable to find information on the height of the near-shore vegetation in Lake Okeechobee, the height of *S. californicus* (giant bulrush) in North America is typically 5-8 feet (USDA, 2003), but the University of Florida Center for Aquatic and Invasive Plants cites heights as tall as 10 feet.[3]

There is a large literature on how emergent plants respond to flooding, although the studies have addressed smaller variations in depth than that which occurs in Lake Okeechobee. In addition, most of the studies have been of a relatively short duration (months), despite Squires and Van Der Valk (1992) identifying that "three years seems to be the minimum time needed for a definitive water-depth tolerance study."

Lentz and Dunson (1998) conducted a greenhouse experiment, during which northeastern bulrush (*Schoenoplectus ancistrochaetus*) was subjected to

---

[3] See https://plants.ifas.ufl.edu/plant-directory/schoenoplectus-californicus/.

varying water levels from −5 to +10 cm (−2 to + 4 in) of the soil surface and attributes of above- and below-ground biomass were examined. In the 9-month experiment, the ratio of below-ground to above-ground biomass decreased with increasing water level, as did the ratio of live to dead shoot biomass. The authors concluded that "even moderate changes in water level may be an important factor [affecting] growth."

Sloey et al. (2016) conducted a similar experiment, with mesocosms to create 0, 40, 60, and 100 percent hydroperiods and to evaluate responses of seedling and adult *S. acutus* and *S. californicus* (the latter being one of the dominant species in Okeechobee) over a 7-month period. They also compared experimental results with findings from field surveys that examined tolerance to different water depths. In the experimental treatment under the longest hydroperiod, where soil oxidation-reduction potentials were lowest and hydrogen sulfide was detected, the survival of *S. acutus* adults was reduced and mortality of the seedlings of both species was 100 percent. Adult *S. californicus* were considerably more tolerant of flooding than *S. acutus*, both in the experiments and in the field. In the experiments, prolonged flooding negatively affected *S. acutus*, while *S. californicus* stems elongated with increased flooding (up to 60 cm or 23.6 in depth). *S. californicus* also was found at the most deeply flooded field sites, where it was noted that the plants maintained their characteristic stem strength, rendering them more tolerant of wave energy. In regard to *S. californicus*, the authors concluded that it "is an excellent species for establishment in high energy tidal environments under a variety of flooding regimes."

Squires and Van Der Valk (1993) identified another factor in addition to shoot morphology that allows *S. californicus* to tolerate prolonged flooding—the species can survive as below-ground tubers for 1-2 years in areas too deep for the plants to grow. Microtopographic variation of lake sediments, which produces different effects of variation in hydroperiod, is also important in the recovery of lake vegetation including *Schoenoplectus* sp. (Nishihiro et al., 2006).

Just one experimental study has been conducted to evaluate effects of prolonged flooding on *S. californicus* collected from Lake Okeechobee, although it too was of short duration (80 days). Smith and Smart (2005) submerged clusters of mature, 3-foot-tall *S. californicus* at various depths in a pond and then measured changes in above-ground and below-ground biomass and stem density. The researchers concluded that "undisturbed bulrush might persist at 3 feet inundation or less, however inundations of greater than 3 feet appear excessive and prolonged periods of greater inundation might cause bulrush stands to fail." The possibility of reemergence of plants from tubers during a subsequent low-water period was not investigated.

This is a snapshot of a large literature on flooding of shoreline emergent plants, with a focus on *Schoenoplectus*, because it has been a focus of SFWMD ecological monitoring in Lake Okeechobee and because its spatial extent is one of the CERP performance measures from the Lake Okeechobee Conceptual Ecosystem Model (Havens and Gawlick, 2005). Research dealing with *S. californicus* supports the view that these plants are well adapted to high-energy shoreline environments, they can tolerate brief periods of total inundation (80 days), and they have approaches to recover from tubers after periods of prolonged flooding. It remains unclear how prolonged flooding at high lake levels might affect the long-term extent and survival of emergent vegetation in the near-shore zone of Lake Okeechobee, which periodically experiences droughts that could allow the plants to recover, both from tubers and by seed germination. A greater understanding of the extent and distribution of viable tubers and seeds could help in predicting the ability of the community to recover from flooding events.

Given the ecological importance of bulrush and co-occurring near-shore plant species in the lake, long-term experimental studies of inundation and the frequency and distribution of stem lengths could provide valuable information to guide planning and inform real-time operational decisions. For example, knowing the frequency distribution of stem lengths and elevations would allow biologists to tell water managers what percentage of the vegetation would be completely submerged at a particular lake depth. A focused monitoring program could also improve the understanding of the time frame for which the plants can tolerate sustained inundation.

Littoral Zone

The diverse mosaic of native vegetation that historically characterized the littoral zone provided an array of ecosystem services (Aumen and Wetzel, 1995). It provided habitat for fish, reptiles, amphibians, migratory birds, and snail kites. The plant assemblage historically was dominated by spikerush, willow, sawgrass, and floating-leaved plants and, in some places, dense surface mats of periphyton. The littoral zone is the most biologically diverse part of the lake, with an estimated 14 bird species, 40 species of adult fish, 35 species of young fish, and 170 invertebrate species along with 30 distinct groups of plants (Havens et al., 1996; Richardson and Harris, 1995).

High lake levels and associated advection of phosphorus into littoral zone areas that normally are nutrient poor could have a variety of adverse effects on the structure and function of the ecosystem (Havens, 2002). These high water impacts include cattail expansion, erosion of the littoral fringe, and impacts to snail kites and wading birds.

Cattail expansion. When lake level rises from 15 to 17 feet, there is an ever-increasing transport of phosphorus from the pelagic zone to the near-shore zone, and eventually into the littoral zone, through natural inlets and boat cuts (Aumen and Wetzel, 1995). Intrusion of phosphorus-rich water results in conversion of areas of diverse plants into cattail monocultures. As in the WCAs, dense cattail monocultures do not provide suitable habitat for fish and birds, and habitat that changes to cattail because of nutrient pollution does not naturally recover to native plants. Hence, the plants of the littoral zone are much less resilient to effects of high water than are the SAV and near-shore emergent vegetation.

Recent vegetation maps (Figure 5-8) suggest that a large part of the littoral zone has undergone a transition to cattail, possibly because of past high water events and influxes of phosphorus. The SFWMD recently eradicated cattail with large treatments of fire and herbicide (the black areas in the second vegetation map). Additional mapping and surveys of birds, fish, and other biota are needed to discern the extent to which native vegetation and its associated ecosystem services return after the treatments. Furthermore, the lake has experienced a substantial increase in the invasive torpedograss *Panicum repens* in the past two decades, and any research and management dealing with cattail also needs to consider this species.

Erosion and berm formation along the littoral fringe. During times of sustained high water levels (near 17 feet), there has been considerable erosion of the edge of the littoral zone and accumulation of large amounts of organic debris along the littoral-nearshore fringe, particularly during high wind events (Havens et al., 2002). The presence of a long-lasting organic berm along the littoral-nearshore interface is a concern because many species of fish in Lake Okeechobee migrate from the littoral to the pelagic zone as they mature (Fry et al., 1999). Blocking the interface could affect the fish assemblage of the lake, the economy they support, and the biota that depend on those fish as food resources. Previously, this organic debris has required removal with heavy construction equipment, which was only possible when a drought left the area dry. One uncertainty is the degree to which the berm formation documented by Havens et al. (2002) was a result of high water alone or high water combined with uprooting of SAV and high wave energy from a hurricane.

Everglades snail kites. One of the critical ecosystem services provided by Lake Okeechobee is habitat for the federally endangered Everglades snail kite. Bennetts and Kitchens (1997) found that the littoral zone of Lake Okeechobee provided a critical habitat of "last refuge" during times of regional drought, when other locations including the WCAs and Kissimmee wetlands were dry. Fletcher

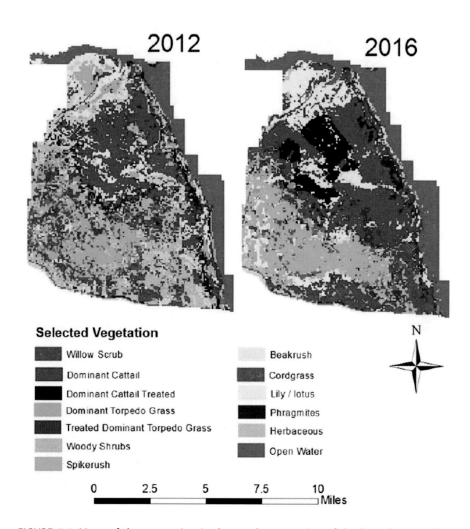

FIGURE 5-8 Maps of the vegetation in the southwest region of the littoral zone of Lake Okeechobee showing a large expansion of cattail from 2012 to 2016. Cattail is shown in red, with treated areas of cattail in black.

SOURCE: Charles Hanlon, SFWMD, personal communication, 2018.

et al. (2017) noted that from 2010 to 2014, the littoral zone of Lake Okeechobee was one of the three most productive wetlands in South Florida for snail kites, contributing to 40 percent of the range-wide fledgling production. The 2016 nesting season in Lake Okeechobee was extensive, with active nests observed from January to November. From a landscape perspective, the littoral zone of Lake Okeechobee is an important central node in a system of regional habitat "modules" (Riechert et al., 2016), allowing for connectivity between a northern and southern component of the regional population network.

Water levels in Lake Okeechobee affect the vegetation structure and its suitability for kite nesting. The eggs of apple snails (the kite's only food source) are laid on the stems of emergent plants, and these eggs die if flooded. The same holds true for kite nests. It is not clear that there is a specific lake level of concern in this case, because the major risk factor is a reversal of lake level (periods when the lake level is declining and then suddenly increases) after kites have constructed their nests above the water surface or after apple snails have laid their eggs. This could happen with a reversal from 14 to 15 feet or a reversal from 15 to 16 feet.

Wading birds. The littoral zone of Lake Okeechobee provides one of the major habitats for wading birds in the regional ecosystem, including heron, ibis, egret, spoonbill, and wood storks (Chastant et al., 2017; Smith et al., 1995). Like snail kites, the littoral zone can serve as a habitat of last refuge for wading birds in years when other areas of the regional ecosystem are dry. However, because wading bird density in Lake Okeechobee is partially a function of the condition in wetlands outside the lake, it remains challenging to fully understand how wading birds respond to variations in water level in Lake Okeechobee.

From January 1988 to September 2002, a comprehensive field study was conducted to quantify the foraging habitats of wading birds in the Lake Okeechobee littoral zone (Smith et al. 1995), including colony turnover, nesting success and productivity, and causes of nest failure (Smith and Collopy, 1995). Wading birds nest in late winter to early spring, when adults forage for small fish to feed fledglings. The study documented that nesting success and foraging both are favored by (1) a spring recession; (2) low to moderate water depths (depths not specified by the authors); and (3) a lack of reversals of water level. As with kites, reversals result in the flooding and loss of wading bird nests, as animals place their nests at a height determined by their perception of future water levels. The researchers provided a specific management recommendation: "moderately high winter lake level [15 feet] followed by a moderate-paced, steady and protracted (5-6 month) recession in water level beginning in December or January." This study led to the development of what has become known as a "spring wading bird window" that continues to be used to evaluate regional planning alternatives. The wading bird researchers also noted that periodically it could be beneficial to have protracted periods of high lake level (15.0 to 15.5 feet) that could allow prey to concentrate, before a year of spring recession.

Chastant et al. (2017) examined empirical relationships between nesting patterns of wading birds and Lake Okeechobee hydrology from 1977 to 1992. This study found that both hydrology and vegetation structure affect nesting success. In particular, the spatial extent of willow (*Salix* spp.) is an important determinant of nesting success. The authors found that nesting numbers and fledgling success were highest with spring water levels falling to the 12- to 13-foot range in a predictable manner. The birds could then establish nests that would be safe from flooding, and the receding water would concentrate prey resources (small fish) in water shallow enough for foraging by both long- and short-legged birds.

MONITORING TO GUIDE OPERATIONAL OPTIMIZATION

From the past research, there is a robust understanding of how certain ecological attributes respond to water level in Lake Okeechobee, yet uncertainty about a number of critical responses remains. There is relatively high certainty that water level reversals negatively affect snail kites and wading birds during the nesting season, and that a receding spring water level into the 12- to 13-foot range supports foraging by birds on small fish. There is relatively high certainty that water levels rising from 15 to 17 feet lead to increasing transport of phosphorus from mid-lake to the near shore and then into the littoral zone, and that the phosphorus can cause cattail expansion. There is also relatively high certainty that multiple years of high water, without intervening lows, lead to reduced spatial extent of SAV. Uncertainties exist regarding the duration of flooding tolerated by the SAV because, as noted earlier, responses depend on the antecedent condition of the plants and the other factors associated with high water.

A comprehensive water quality and ecological monitoring program exists for Lake Okeechobee, but there may be opportunities to modify it in ways that increase the value of the information collected for decision making. For example, the sampling of SAV includes a yearly mapping program to discern the spatial extent at the end of the summer season, and it includes quarterly sampling of transects in the region where SAV is known to occur. If that quarterly sampling were replaced by more frequent sampling of SAV at a smaller number of sentinel sites, data on the recent condition of the SAV (e.g., robust, recovering) could be used to predict the likely impacts to SAV of holding more water in the lake at that time and could inform real-time water management decisions regarding lake management. Over time, continued data collection would test the validity of the prediction, building the knowledge base about effects of depth on this plant assemblage. Targeted monitoring could also enhance understanding of the effects of high water conditions on plants and animals in the lake, thereby informing the lake regulation planning process and improving the performance of the Lake Okeechobee environment model.

MODELING

In the past, evaluation of lake regulation schedules has depended largely on regional hydrologic model results and performance measures for select ecological attributes in the lake that are based on past research. Because of the uncertainties described above, there are shortcomings to this approach. Those uncertainties could be reduced by integrating a recently developed and validated lake ecosystem model (Jin and Ji, 2013) into the process. Furthermore, this same model might be used in real time for operational optimization while managing the lake under a particular regulation schedule.

The Lake Okeechobee Environment Model (LOEM) is a coupled hydrodynamic-wind wave-sediment resuspension and transport model with more than 2,100 grid cells and five vertical layers (Jin and Ji, 2001, 2013). The model has a submerged vegetation component that considers wave energy, water depth, turbidity, and plant growth rates in response to light attenuation and sediment nutrients. The LOEM is a tremendous advance over previously used models that treated the lake as one unit, rather than examining vertical and horizontal variability, which today is the norm for ecosystem models. Therefore, it can model sediment resuspension and transport around the lake, as well as phytoplankton and plant densities in particular locations of the lake. The LOEM has effectively predicted the lake-wide spatial extent of SAV and the temporal dynamics (acreage and biomass) at a particular sampling location over a 9-year time period. Because the model can also predict the transport of suspended solids and phosphorus within the lake and the wave energy on the western shoreline, it could be used to predict how a particular water level regime might influence a variety of conditions, including SAV spatial extent, erosion potential at the littoral fringe, and phosphorus movement into the littoral zone. If an empirical relationship can be derived from historical monitoring data, it may also be possible to use the model to predict cattail expansion.

In summary, the SFWMD now has a sophisticated modeling tool to help screen alternatives in a regulation schedule review. The tool might also play a role in projecting how certain key attributes of the lake might respond to future changes in water level as a part of operational optimization. This lake model will need to be used in concert with regional models to evaluate systemwide benefits and tradeoffs of different approaches to manage the lake and other parts of the broader ecosystem.

TRADEOFFS

The current regulation schedule for Lake Okeechobee holds water at a considerably lower level than the prior schedule. Although this provides potential benefits for nearly every in-lake ecological attribute discussed in this chapter, it may provide widespread negative effects for Everglades dry season flows, the Northern Estuaries, and water supply. Those negative effects happen because of the tremendous loss of regional water storage compared to the earlier lake regulation schedule.

When a new Lake Okeechobee regulation schedule is considered, the analysis of alternatives performed by the USACE must consider the tradeoffs that exist between potential impacts to the condition and ecosystem services of the lake, the availability of water for human uses, and potential benefits to the condition and ecosystem services of downstream ecosystems (i.e., the Northern Estuaries and the remnant Everglades). Regulation schedule evaluations always are performed in this broad regional context. The challenge is to perform a systemwide analysis of alternatives that allows water managers to select a lake regulation schedule that maximizes the benefits, minimizes the adverse impacts where possible, and balances the tradeoffs, considering the latest science and real-time operational capabilities.

CONCLUSIONS AND RECOMMENDATIONS

Lake Okeechobee is the last major component of water storage in the northern end of the South Florida ecosystem to be resolved, and its regulation schedule has significant implications for conditions throughout the ecosystem. The lake regulation schedule will soon be revisited to determine new operational rules. The completion of the Herbert Hoover Dike rehabilitation project could enable higher water levels to be held within Lake Okeechobee, although the feasibility of higher water levels must still be determined through an updated risk assessment. The regulation schedule revision process also considers tradeoffs among the ecological conditions in the lake, the Northern Estuaries, and the Everglades, as well as water supply and flood management. Hydrologic and ecological modeling tools have been developed to assess potential benefits and impacts from various regulation schedules on the lake and broader region. To inform that process and in response to frequent questions about the impacts of increased water levels on the ecology of Lake Okeechobee, the committee summarized the latest information and identified key research needs to help inform the within-lake portion of the tradeoff analysis.

The magnitude of ecological impacts in the lake from additional storage will depend upon antecedent ecological conditions. Improved understanding of these dependencies could be used to inform real-time operations to reduce adverse ecological effects and provide more flexibility given appropriate risk tolerance in lake management. A new regulatory schedule that stores more water in Lake Okeechobee would require tradeoffs between in-lake ecological impacts and ecological and water supply benefits throughout the South Florida ecosystem. Past research has shown that ecological conditions in the lake are adversely affected by high water levels (above ~16 feet) and multiple consecutive years without low water levels (~12 feet). Additionally, reversals of water level recession during spring nesting can adversely affect wading birds and snail kites. However, there are considerable uncertainties about high water impacts to SAV and near-shore emergent vegetation, which provide important ecological services in the lake, because many of the effects of high water depend on antecedent conditions. For example, high stage effects on SAV vary depending upon whether the plants are healthy and mature, stressed, or just recovering after a prior impact. Reducing those uncertainties and using that information to inform operations could reduce the ecological impacts associated with increased storage.

Adjustments to Lake Okeechobee monitoring and full integration of modeling tools would provide rigorous science-based information to support a regulation schedule review and real-time optimization of operations under any regulation schedule. Refinements to the ecological monitoring and adaptive management program could reduce critical uncertainties, inform lake regulation schedule planning, and enhance real-time lake operations. For example, as discussed earlier in this chapter, moving from quarterly transect sampling of SAV to more frequent sampling at just a few representative sites might provide more actionable information and lead to a better understanding of the effects of antecedent conditions. Monitoring could also improve the understanding of the potential impacts from inundation to emergent vegetation in the near-shore zone. Further, the Lake Okeechobee Environment Model is a tool to use in concert with regional hydrologic and ecological models to evaluate the implications of alternative regulation schedules and lake operations, particularly as new data become available to refine the model's SAV component.

6

A CERP Mid-Course Assessment: Supporting Sound Decision Making for the Future Everglades

A core theme of the committee's 2016 report (NASEM, 2016) was the critical need for a forward-looking, systemwide analysis to reexamine restoration outcomes and Comprehensive Everglades Restoration Plan (CERP) goals, objectives, and components in light of recent and potential future changes. The National Academies of Sciences, Engineering, and Medicine (NASEM, 2016) noted several key issues that had emerged since the CERP's inception in 2000 that were each likely to have significant, systemwide impacts on the outcomes of restoration efforts—advancements in scientific and engineering knowledge related to the understanding of pre-drainage hydrology, climate change and sea-level rise, and the feasibility of storage alternatives.

Just such a need was anticipated in the original CERP Programmatic Regulations (33 CFR §385.31), which call for regular 5-year comprehensive assessments of the program and progress anticipated in light of new information and understandings, termed "CERP updates." An initial CERP update was completed in 2005 but was not revisited in the decade to follow. NASEM (2016) recommended that an assessment of systemwide CERP benefits be completed in conjunction with program-level adaptive management to ensure that the CERP is based on the latest scientific and engineering knowledge, considers long-term ecosystem needs, addresses potential restoration conflicts, and is robust to changing conditions. Such an effort would better inform current and future project and systemwide program planning efforts and would assure decision makers and the public that, nearly two decades after inception, the CERP is still on track and the best restoration investments are being pursued.

CERP agencies have not acted on the NASEM (2016) recommendation for a CERP systemwide assessment for a variety of reasons, including that (1) an update is not needed because new knowledge has already been incorporated into each individual project planning effort and (2) undertaking a CERP update would require reassigning limited staff and resources, thereby slowing the momentum of current CERP planning and implementation efforts. This committee is specifically charged to report to Congress not only on progress made but also on scientific and engineering issues that may impact progress (see task in Chapter 1). The committee remains unconvinced that the current, individual, project-level planning approach is an effective means of reassessing the systemwide, scientific-guiding vision for restoration in light of the extended expected time frame for completing the CERP, changing system conditions, and the evolving understanding of the future Everglades ecosystem. As discussed in Chapter 3, no recent individual project planning effort captures the systemwide outcomes of the CERP projects concurrently in planning, and most projects have failed to assess project performance under changing future climate conditions.

It is critically important that the CERP be robust across a large range of temperature, rainfall, sea level, and population regimes that may drive the system as restoration is completed. The original CERP was formulated based on a pre-drainage vision of the historical Everglades and the assumption that rainfall and temperature time series observed during the 1965-1999 period captured the full range of variability that would have been observed under pre-drainage conditions as well as that expected throughout the 21st century. There is now ample evidence that rainfall and temperature distributions in South Florida historically have exhibited multidecadal variations outside the 1965-1999 (or updated 1965-2005) period of record (Enfield et al., 2001; SFWMD, 2011). There is general consensus among climate projections that average temperatures in South Florida will increase over time because of increases in atmospheric greenhouse gases, but considerable uncertainty about future rainfall patterns remains (Carter et al., 2014; Dessalegne et al., 2016; Irizarry et al., 2013; Misra and DiNapoli, 2013; Misra et al., 2012a). There is compelling recent evidence that sea-level rise in South Florida is accelerating and expected to continue in the future (NOAA, 2017). These changes will have profound impacts on the South Florida ecosystem and the ability to provide flood protection and meet the water and recreational demands of a growing population.

Florida continues to be one of the fasting growing states, and it has recently passed New York as the third most populous state. Florida's population was approximately 16 million when the Yellow Book was completed; it is projected to grow to approximately 21.5 million by 2020 and to more than 26 million by 2040 (Rayer and Wang, 2018). Growth at this rate (nearly 700 people per day) will continue to exert development pressures in South Florida. Volk et al. (2017) project that, at current trends, total developed land in Florida could increase from 6.4 million acres in 2010 to 11.6 million acres in 2070, representing an additional conversion of approximately 14 percent of the total land area in Florida. Future population growth and development has important implications for land and water use and will add to the challenges associated with flood management and water quality.

The committee is sympathetic to the concerns about the opportunity costs associated with reassigning limited staff and resources. NASEM (2016) was complimentary of the pace of restoration and tried to make clear that the recommendation for a systemwide CERP assessment was neither a call for "pencils down" nor for an overhaul of the CERP itself. The committee remains impressed by, and supportive of, the current pace of construction and project planning efforts and expects the agencies to continue CERP implementation efforts while a systemwide CERP assessment is pursued. By mid-2019, tentatively selected plans will have been developed for all of the major central CERP storage projects east, west, south, and north of Lake Okeechobee, with the exception of the Lake Belt in-ground reservoirs. Now that the vision for CERP storage is largely developed and that CERP authorized and soon-to-be-authorized projects will require decades to construct at current funding levels, the time is right to undertake a mid-course assessment.

The committee understands that a mid-course assessment of the CERP might provide information that could motivate a significant recalibration of the original restoration goals. This evolution of the CERP is exactly what was envisioned when the CERP was launched within an adaptive management framework. The mid-course assessment could inform optimal final designs and integration of the individual projects. A systemwide assessment is also essential to ensure that the program-level adaptive management uncertainties that RECOVER identified as "showstoppers" (RECOVER, 2015) are addressed in a timely way, that the CERP is designed for expected future conditions, and that critical transitions can be anticipated, planned for, and more effectively managed. Such an assessment could also inform potentially complementary efforts such as the Southeast Coastal Assessment[1] focused on sea-level rise and coastal vulnerability.

The Everglades of 2050 and beyond will differ from what was envisioned at the time of the Yellow Book. Thus, despite the expressed agency concerns, this committee remains fully supportive of the NASEM (2016) recommendations and the importance of forward-looking program-level analysis that incorporates the latest socioeconomic, scientific, and engineering information, while considering uncertainties about future conditions. The committee notes that even a $10 million investment in such assessment would represent only 0.05 percent of what is likely to be at least a $20 billion restoration effort. This outlay would seem prudent, to ensure that the guiding programmatic vision for restoration as well as future project planning effectively incorporates current knowledge and changing system conditions. This systemwide analysis would also assure the public that scarce public funds are being invested in a manner that maximizes future restoration benefits.

---

[1] See http://www.sad.usace.army.mil/SCA/.

In the balance of this chapter, the committee presents new information on sea-level rise and storage that further underscores the need for systemwide analysis; provides guidance on the types of systemwide analysis envisioned for a mid-course assessment; identifies critical research needs to better support CERP planning and implementation in light of future stressors; and suggests programmatic changes that could provide for more effective integration and use of science to inform decision making.

UNDERSTANDING THE CHANGES AFFECTING THE CERP

The Everglades ecosystem has changed dramatically in the last 100 years. While restoration efforts seek to regain characteristics of the historic Everglades ecosystem and support productive fish and wildlife habitat, external drivers such as climate change, species invasions, sea-level rise, land-use changes, and water use influence the ability to achieve this outcome. These internal and external forces on South Florida lead to an ever-changing mosaic of human and natural system elements that impact the outcomes of the CERP. This section presents recent information on two areas of change affecting the CERP—climate change and advances in understanding CERP storage—and their potential implications to restoration planning.

Understanding Climate Change and Sea Level Rise

Many aspects of global climate are changing (USGCRP, 2017) and have implications for the South Florida ecosystem, including changes in surface, atmospheric, and oceanic temperatures; rising sea levels; and ocean acidification. Past Committee on Independent Scientific Review of Everglades Restoration Progress (CISRERP) reports have discussed the possible effects of changes in precipitation (including interannual and seasonal variability) and increasing evapotranspiration on Everglades water budgets (NASEM, 2016; NRC, 2014). This section focuses on new understanding of the implications of sea-level rise on restoration outcomes, based largely on new information since the publication of NASEM (2016). One of the most prominent features of South Florida, and one of its key vulnerabilities in the face of continued sea-level rise, is its 3,400 mi^2 (8,750 km^2) of land area situated below 5 ft (1.5 m) elevation (Titus and Richman, 2001). Sea-level change can cause a number of impacts in coastal and estuarine zones, including inundation or exposure of low-lying coastal areas, changes in storm and flood damages, shifts in extent and distribution of wetlands and other coastal habitats, changes to groundwater levels, and alterations to salinity intrusion into estuaries and groundwater systems (CCSP, 2009).

Rates of sea-level rise have been accelerating recently, from a long-term global mean of 1.5-1.9 mm/yr (1920-2016) to 3.3±0.4 mm/yr (1993-2016). Higher acceleration rates have been recorded in Florida, with long-term rates of 2.4±0.1 (1920-2016) increasing to 7.6±1.3 mm/yr (2000-2016) as observed in Key West (Valle-Levinson et al., 2017). The Southeast Florida Climate Compact (2015) developed unified sea level–rise projections (Figure 6-1), ranging from a scenario of 2.6 ft (0.8 m) by 2100 (the IPCC [2014] median scenario) to 6.8 ft (2 m) by 2100 (NOAA [2014] "high risk" estimate). Incorporation of updated sea-level rise considerations into large-scale ecosystem restoration planning could have substantial implications for planned restoration actions. For example, using updated sea-level rise scenarios in the 2017 Louisiana Coastal Master Plan, compared to the 2012 plan, resulted in dramatically different predictions

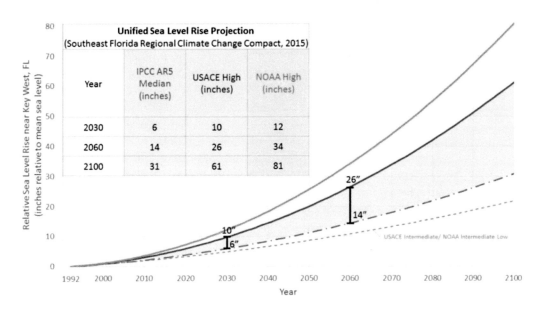

FIGURE 6-1 Unified sea-level rise projection, referenced to the Key West tide gauge.

NOTE: The Southeast Florida Regional Climate Change Compact based its projections on three global curves: the median of the International Panel on Climate Change (IPCC) Assessment Report 5 (AR5) Representative Concentration Pathway (RCP) 8.5 scenario (blue dashed curve), which represents the lowest boundary; the USACE High projection representing the upper boundary until 2060 (solid blue line), and the NOAA High curve representing the uppermost boundary for medium- and long-term use (orange solid curve). The USACE Intermediate or NOAA Intermediate Low curve is also shown for comparison (green dashed curve).

SOURCE: Southeast Florida Regional Climate Change Compact (2015).

of the state of the future coastal landscape and the magnitude of future coastal storm damages (CPRA, 2012, 2017). The California Ocean Protection Council has recently updated guidance for incorporating sea-level rise projections into planning, design, construction, and other decisions (COPC, 2018).

Sea-level rise interacts with other effects of the changing climate including freshwater availability, increasing temperatures, and acidification. Such future changes challenge both the human system and the natural system. The following sections present three examples of how sea-level change and other climate factors can result in transitions within the natural system that may influence CERP outcomes, focusing on effects on wetland peat, northern estuaries, and Florida Bay. How the effects of sea-level rise interact with restoration actions and other system changes has important implications for the future of the ecosystem.

Sea-Level Rise Effects on Wetland Peat

The degree to which sea-level rise results in wetland loss is a complex but critically important issue for South Florida and, more broadly, coastal wetlands globally. The coastal wetland landscape can respond to sea-level rise in three potential ways: (1) peat and sediment accretion that allows coastal wetlands to keep pace with sea-level rise, (2) submergence with landward migration of coastal vegetation and wetland habitat, or (3) submergence and loss of wetland ecosystem habitat, without habitat migration (Chambers et al., 2015). A critical question for the CERP is whether, and for how long, restoration of freshwater flows can mitigate salinity incursion related to sea-level rise and associated peat collapse and can facilitate a landward migration of coastal mangroves to counteract the effects of sea-level rise.

The response of coastal wetlands to sea-level rise depends on a number of counteracting factors. Accretion can occur by the accumulation of mineral sediments and organic matter (Day et al., 2000). The importance of these soil components varies among wetland types and with their proximity to sources of sediment. With sea-level rise, increased frequency and depth of tidal inundation could potentially increase transport of mineral matter to fringe mangroves. The degree to which mineral sediments are deposited in wetlands decreases with distance from the coast or freshwater inflows that are sources of sediment. Sometimes more important than regular tidal or riverine supply of sediment are the episodic inputs from hurricanes. For example, in 2005, sediment deposition to Shark River Slough from Hurricane Wilma was 0.5-4.5 cm, many times greater than annual accretion rates (Castañeda-Moya et al., 2010). For large areas of the coastal Everglades, however, sediment supply is relatively low, so organic matter dynamics largely drive rates of accretion (Chambers et al., 2015). Vertical accretion of organic matter is the net response of above-ground and below-ground plant production, decomposition, inputs of sediment-bound organic matter deposited on the wetland surface, and changes in soil and root bulk density.

A meta-analysis of global data for mangroves indicated that 80 percent were accreting at a rate equal to or exceeding sea-level rise (Alongi, 2008). This pattern is consistent with research findings from the coastal fringe mangroves of Shark River Slough, where rates of accretion from 1924 to 2009 (2.5-3.6 mm/yr) exceed long-term rates of sea-level rise (2.2 mm/yr) (Smoak et al., 2013), but probably not the recent increases to 7.6±1.3 mm/yr (2000-2016) as observed in Key West. However, there is likely considerable spatial variability in landscape response to sea-level rise.

The dynamics of processes that accelerate or retard vertical accretion of organic matter in the coastal Everglades under changing conditions remain incompletely understood and potentially involve a host of mechanisms. For example, increases in tidal inundation and salinity penetration could promote accretion when salt stress reduces microbial action and reducing conditions decrease aerobic decomposition (Chambers et al., 2013, 2014). In addition, because the Everglades is phosphorus-limited, increased supply of phosphorus from Florida Bay associated with sea-level rise and coastal storm surge events could stimulate plant and periphyton production and therefore the accumulation of soil organic matter (Childers et al., 2006; Rivera-Monroy et al., 2007). On the other hand, increases in salinity could facilitate the net loss of soil organic matter through several possible mechanisms. Enhanced decomposition of soil organic matter can occur through an increased supply of the terminal electron acceptor sulfate from sea water, which inhibits methanogenesis and/or shifts the dominant pathway of decomposition toward sulfate reduction (Chambers et al., 2011; Neubauer et al., 2013). Additionally, salt or sulfide generated through enhanced sulfate reduction may cause stress to vegetation, which diminishes above-ground and below-ground production (Batzer and Shartiz, 2006; Castañeda-Moya et al., 2011, 2013; Troxler et al., 2013). Of particular concern for Everglades restoration, saltwater intrusion and increased inundation can cause plant mortality and the collapse of root structures, resulting in subsidence and greatly diminishing the integrity of peat soils. Investigators have also reported decreases in the bulk density of peat soils (Chambers et al., 2014) or loss of root turgor (DeLaune et al., 1994) associated with saltwater inundation, which can contribute to the "collapse" of the peat soils into open water with the sudden loss of elevation and death of wetlands plants. Herbert et al. (2015) describe this as an alternative stable state—freshwater/brackish communities die back, roots die, and there is the structural collapse of peat to open water before saltwater vegetation can reestablish.

The low gradients of much of the South Florida coast enable landward mangrove migration in response to sea-level rise more than in many other coastal areas (Spalding et al., 2014). Ideally during sea-level rise, as salinity penetrates further inland, up-slope freshwater marshes give way to mangroves. Work by Ross et al. (2000) has shown a 50-year vegetation composition shift in the marsh areas of the Southeast Saline Everglades with sea-level rise. This increase in salt-tolerant species such as red mangrove shows that gradual shift can occur in response to modest rates of sea-level rise. However, if increased salinity due to sea-level rise or storm surges impairs salt-intolerant vegetation and compromises the integrity of soil at a rate that exceeds the ability of salt-tolerant vegetation to occupy this space, then peat collapse and ponding can occur in the freshwater wetlands (Chambers et al., 2015; Wanless and Vlaswinkel, 2005).

There is evidence of peat collapse in sawgrass wetlands in South Florida where increased salinity due to sea-level rise or storm surge stresses freshwater vegetation at the upper edge of the coastal ecotone (Figure 6-2). Ongoing experimental studies in Everglades National Park and in controlled mesocosms (Figure 6-3) show the combined influence of salt addition and hydroperiod affect plant production, net ecosystem exchange, and porewater chemistry—all influencing the carbon balance and, ultimately, the peat soil stability of these ecosystems (Mazzei et al., 2018; Wilson et al., 2018; T. Troxler, FIU, personal communication, 2018). Understanding the rates at which these processes occur and key thresholds of salinity and hydroperiod is crucial to predicting future conditions in South Florida. Unless salt-tolerant vegetation can migrate to stabilize these zones, without roots to maintain soil structure and limited carbon production coupled with accelerated decomposition, organic peat collapses and becomes transformed into ponded areas (Figure 6-2). Once ponding occurs, it may be difficult for mangroves to reclaim these areas if the water level becomes too deep to allow for colonization of mangrove propagules or limits dispersal mechanisms. Planting of mangrove propagules to accelerate colonization or to compensate for limited dispersal opportunities in existing vegetation could be warranted.

Freshwater withdrawals or seepage from the remnant Everglades will likely accelerate the potential for peat collapse due to salinity incursion, while increased freshwater deliveries may be able to offset the effects. Meeder et al. (2017) found that during the past century sea-level rise was accompanied by saltwater encroachment, which was controlled by the elevation of high tide and varied widely among the five watersheds studied because of differences in freshwater discharge. In only one of the watersheds was freshwater supply adequate to maintain a plant community resulting in a more rapid rate of sediment accumulation than the other sites. Under diminishing freshwater discharges

FIGURE 6-2 Aerial view of open water "pothole" ponds spread throughout a sawgrass marsh surrounded by mangroves in northwest Cape Sable, Everglades National Park. These ponds are thought to have been formed through collapsing peat driven by saltwater intrusion.

SOURCE: Steve Davis, Everglades Foundation.

FIGURE 6-3 Experimental plots established in a collapsing brackish water sawgrass marsh. Scientists are investigating how saltwater may influence peat collapse.

SOURCE: Ben Wilson, FIU.

and increasing sea-level rise, Meeder et al. (2017) see little hope to mitigate loss of the Southeast Saline Everglades over the long term. Recent work by Dessu et al. (2018) in Shark River Slough found rising magnitude, frequency, and duration of salinity in the coastal sites, as well as seasonal patterns with greater salinity during the dry season. The study points to the need for increasing flows in Shark River Slough, with particular attention during the dry season to reduce salinity intrusion and mitigate peat collapse. It is not clear, however, how long restoration actions that increase flow will be able to mitigate against salinity intrusion and peat collapse as seas continue to rise. As sea level continues to rise in South Florida, its threat to coastal wetlands becomes more profound. Tools such as the Marsh to Ocean Index (Park et al., 2017) may be useful to represent large-scale patterns of change, but management measures must be grounded in the processes of ecosystem change. Interim and overall restoration goals should reflect what can reasonably be accomplished in the face of these larger regional changes.

Sea-Level Rise and Estuary Restoration Goals

Certain expectations about coastal ecosystem restoration for estuarine fauna may require modification as sea-level rise results in salinity that exceeds the tolerance of some estuarine organisms. At certain locations, it may not be possible to combat high salinity with increased freshwater flow from CERP projects.

Oyster reef restoration serves as an example. One goal of the CERP is to enhance the spatial extent of oyster reefs in the Caloosahatchee Estuary, St. Lucie Estuary, Loxahatchee Estuary, and Lake Worth Lagoon. Oyster reefs provide essential habitat for fish, crustaceans, mollusks, worms, and other biota (Volety et al., 2009). Oysters also filter particles from the water and can have a positive influence on water quality (Coen et al., 1999). One goal of the CERP is to reduce the occurrence of prolonged low-salinity events in estuaries, caused by large freshwater discharges, because past events have killed entire oyster populations (Volety and Tolley, 2005; Volety et al., 2009). However, despite projections, the effects of sea-level rise on northern estuary oyster populations or restoration outcomes have not been adequately examined.

Effects of high salinity have recently been studied in another Florida estuary—the Apalachicola (Camp et al., 2015; Huang et al., 2015). During low rainfall periods, future salinity is projected to be more favorable for marine predators and pathogens of oysters, including worms, sponges, gastropods, and internal unicellular parasites (Camp et al., 2015).

The CERP Monitoring and Assessment Program (RECOVER, 2009) includes oyster monitoring in locations that are likely to have oceanic salinity levels and therefore adverse conditions for oysters at the projected 2060 sea level. Particularly at risk are the Tarpon Bay and Bird Island sites in the Caloosahatchee Estuary (Figure 6-4), sampling sites 1-3 in the St. Lucie Estuary, and the entire Lake Worth Lagoon (Figure 6-5). In contrast, increased salinity might counterbalance effects of freshwater runoff in the north and south forks of the St. Lucie Estuary, as well as sites farther up the river in the Caloosahatchee, and thereby create conditions more favorable to oyster growth.

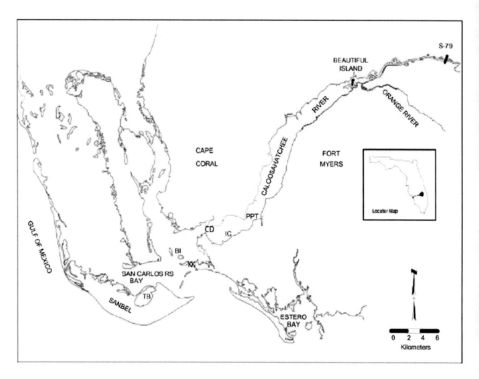

FIGURE 6-4 CERP oyster sampling sites in the Caloosahatchee Estuary.

SOURCE: Volety et al., 2009.

Sea-level rise will affect the ability of certain locations in the South Florida ecosystem to continue to support oyster reefs. Analysis and assessment are needed to predict the magnitude of effects, and the results should be used to inform expectations regarding oyster growth and survival under restoration conditions. In certain places, such as the Lake Worth Lagoon and the main bays of the St. Lucie and Caloosahatchee Estuaries, it may be unrealistic to set specific goals for any oyster reefs in future decades, regardless of what is done to restore freshwater flow. In the Lake Worth Lagoon, there is no place for oysters to migrate when salinity becomes too high to allow them to survive. In other places, such as the St. Lucie Estuary, oysters might migrate upstream with rising estuarine salinity, assuming that there is suitable substrate and a source of larvae. CERP agencies should consider whether long-term monitoring locations should be changed over time so that the data reflect the health of reefs, wherever they occur, in any given decade.

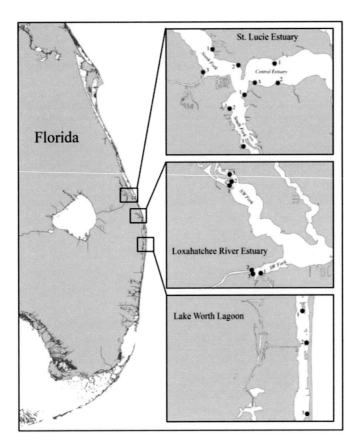

FIGURE 6-5 CERP oyster sampling sites in southeast Florida.

SOURCE: FWC, 2015.

Florida Bay Restoration and Climate Change

Florida Bay, at the southern end of the Everglades, is a large (850 mi^2 [2,200 km^2]) semi-enclosed shallow embayment internally divided by a series of carbonate mudbanks. These banks restrict ocean flushing and create a series of basins with variable residence time, salinity, and biogeochemical character. The shallow depth, averaging about 3 ft, enables light penetration to support extensive seagrass beds. Evidence suggests that South Florida estuarine geology and biota are sensitive to changes in climate, with ecological regime shifts across Florida Bay and Biscayne Bay in the mid-1950s and early-1960s and across the southwest coastal margin in the 1980s (Wachnicka and Wingard, 2015; Wingard et al., 2017).

Given that the freshwater budget of Florida Bay is dominated by exchanges with the atmosphere rather than inputs from runoff (Nuttle et al., 2000), there is potential for future changes in evaporation or precipitation to influence salinity and thus ecological change. It is hypothesized that the combined effect of increased temperature and salinity caused the seagrass die-off and subsequent phytoplankton blooms in the 1980s and early 1990s (Koch et al., 2007). Higher temperatures and potential increased risk of long-term drought associated with climate change could pose additional future risks to Florida Bay. However, such changes due to climate should be considered in the light of sea-level rise, which may increase oceanic connectivity if mudbank elevation becomes relatively lower, increasing flushing and potentially mitigating stress to seagrasses (Koch et al., 2015). Future changes in bay functioning with sea-level rise and climate change are likely to vary across the bay because of the differential ability of mudbanks to adjust to changing conditions (Wanless and Taggett, 1989) and interactions with localized inflows from, for example, Taylor Slough.

Climate change studies of Florida Bay habitat suitability for fishes and invertebrates, considering only salinity and temperature variables, found that "the estuarine fauna of Florida Bay may not be as vulnerable to climate change as other components of the ecosystem" (Kearney et al., 2015). The study did show that "temperature increases alone negatively affected the availability of optimal habitat for all species, except that of juvenile spotted seatrout," but the habitat suitability approach does not consider movement of species to areas with better conditions (which could be easier if mudbanks become relatively lower due to sea-level rise). Kearney et al. (2015) acknowledged that their study does not consider the broader ecosystem effects of climate change on Florida Bay. Using models that are limited in their ability to encompass the complex system dynamics may give a false sense of security regarding future change.

These studies highlight the potential sensitivities of Florida Bay to climate change, sea-level rise, and other factors, such as the rates of carbonate accretion in the mudbanks relative to sea-level rise. The potential geomorphic change influencing bay connectivity needs to be better understood to grasp the interacting implications of climate change and sea-level rise for the effectiveness of CERP-planned actions. Such analysis should be conducted with a system-level view, accounting for potential changes in Florida Bay inflows under a range of future climate conditions to understand the capacity of the CERP to improve the resilience of the Southern Estuaries.

Understanding Changes in CERP Storage

The key to improving the condition of the South Florida estuaries and remnant Everglades ecosystem is creation of storage and conveyance projects that both reduce the amount of water lost to diversions and facilitate temporal and spatial patterns of releases that more closely resemble the pre-drainage system. The CERP, as authorized by Congress in 2000, included several conventional surface reservoirs totaling more than 1 million acre-feet (AF) of capacity, three in-ground reservoirs with more than 300,000 AF, and 333 aquifer storage and retrieval (ASR) wells with an effective capacity of more than 4 million AF (see Table 6-1).

Projects originally included in the CERP required more detailed investigations about their feasibility and design, as well as authorization and appropriations by Congress prior to initiation of construction. During the time it took to undergo those processes, certain CERP projects were substantially reduced in magnitude (Table 6-1). For example, storage north of Lake Okeechobee has been reduced from 200,000 AF of above-ground storage and 200 ASR wells to a 43,000 AF reservoir and 80 ASR wells (USACE and SFWMD, 2018c). The Regional ASR Study (USACE and SFWMD, 2015b), a large-scale pilot study, concluded that only approximately 131 wells could be constructed without impacting the water supply of other users. Reduction of ASR storage capacity in the CERP by two-thirds would reduce all planned storage in CERP by about 50 percent. Plans now include a 240,000 AF reservoir in the Central Everglades Planning Project for the EAA Storage Reservoir (SFWMD, 2018a). The feasibility of two in-ground reservoirs near Everglades National Park, the North and Central Lake Belt Projects that would have added approximately 280,000 AF of storage, has also been questioned, and very little, if any, progress has been made to resolve uncertainty about the Lake Belt reservoirs.

As the largest surface-water storage component in the South Florida ecosystem, Lake Okeechobee is a critical component of regional storage. NASEM (2016) noted that changes in the Lake Okeechobee operating schedule to protect the Herbert Hoover Dike during repairs have dramatically reduced regional storage (as much as 480,000 to 800,000 AF) compared to the original CERP planning assumptions. The agencies plan to revisit the operating policy for Lake Okeechobee (currently scheduled for 2019), but it is not known what portion, if any, of that lost storage will be regained by the adoption of a new schedule, which will also consider adverse effects on increased water levels on the lake ecosystem (see Chapter 5).

Overall, these represent substantial reductions in storage compared to that proposed in the CERP, but the implications to CERP outcomes systemwide has not been examined. Recent sensitivity modeling for the EAA Storage Reservoir

TABLE 6-1 Proposed and Updated Capacities of Storage Components of the Restoration Plan

STORAGE COMPONENT	Yellow Book Storage Capacity Acre-Feet	Updated Storage Capacity Acre-Feet
Existing System Storage		
Lake Okeechobee	3,817,000[a]	3,253,000[a]
Water Conservation Areas	1,882,000	1,882,000
Total lake/WCA storage	**5,699,000**	**5,135,000**
Above-ground Reservoirs		
North Storage Reservoir (Kissimmee)	200,000	43,000[b]
Taylor Creek/Nubbin Slough	50,000	0[b]
Caloosahatchee (C-43) Basin	160,000	170,000
C-44 Reservoir	40,000	50,600
Other Upper East Coast Storage[c]	349,000	109,400[c]
EAA Reservoirs	360,000	300,000[d,e]
Central Palm Beach Reservoir	19,920	TBD
Site 1 Reservoir	14,760	0[f]
Bird Drive Reservoir	11,600	0[g]
Acme Basin	4,950	0[h]
Seminole Tribe Big Cypress	7,440	TBD
Total above-ground reservoir storage	**1,217,670**	
Projects planned to date	1,190,310	673,000
Potential storage in projects not yet planned, or planning not finalized:	27,360	
In-ground Reservoirs		
North Lake Belt	90,000	Feasibility unproven
Central Lake Belt	187,200	Feasibility unproven
L-8 Basin	48,000	45,000[e]
Total in-ground reservoir storage	**325,200**	
Projects planned to date	48,000	45,000[e]
Potential storage in projects not yet planned, or planning not finalized:	277,200	
ASR Wells		
All CERP wells	1,637,000[i]	TBD

[a] Updated capacity based on difference between an assumed low level of 9 ft and the highest stage in the upper band (17.25 ft for Lake Okeechobee Regulation Schedule (LORS) 2008 and 18.5 ft for Water Supply and Environment [WSE]), based on calculator in http://www.sfwmd.gov/gisapps/losac/sfwmd.asp based on the polynomial model. This capacity may change based on a planned update to the LORS.

TABLE 6-1 Continued

[b] Based on tentatively selected plan for the Lake Okeechobee Watershed Restoration Project; planning process ongoing.
[c] Includes C-23, C-24, C-25, and St. Lucie North and South Fork reservoirs and natural storage areas. Although the storage capacity decreased significantly between the original CERP framework and the final Indian River Lagoon South project implementation report modeling analyses showed that the CERP objectives for the IRL-S project could be reached with substantially less storage.
[d] Includes EAA Reservoir and A-1 FEB, which was constructed for Restoration Strategies.
[e] An FEB is operated with the primary objective to optimize performance of the STAs (e.g., reduce excessive loading and periods of drydown) rather than to optimize the quantity or timing of water flow to the natural system. Therefore, the hydrologic benefits may be less than other storage features, depending on their operational plans and objectives.
[f] The Site 1 Impoundment plan would provide 13,280 AF if constructed, but the SFWMD) has proposed not completing the reservoir.
[g] The project delivery team determined this project to be infeasible.
[h] Land sold before it could be acquired. Remaining project elements completed outside of the CERP.
[i] Maximum annual storage determined by maximum annual inflows minus maximum annual outflows, over the period of record. Eighty wells are proposed for the Lake Okeechobee Watershed Project, but the maximum storage provided by these wells was not available.

SOURCES: USACE and SFWMD (1999, 2004b, 2010, 2014) and NRC (2005).

as part of the CEPP post-authorization change report (see Chapter 3) suggests that less total storage may be needed than originally envisioned in the CERP to achieve the original objectives for average annual flow into the Everglades from the northern boundary and for reductions in high water discharges from Lake Okeechobee to the Northern Estuaries (SFWMD, 2018a). This effort highlights the importance of analyzing the combined effects of all projects, informed by improved modeling tools and operational strategies, to understand the systemwide outcomes from authorized and planned CERP projects. By the end of 2019, the planning for most major storage components will be complete, with only Lake Okeechobee Regulation and the Lake Belt projects unresolved.

SPECIFIC ANALYSES NEEDED FOR THE MID-COURSE ASSESSMENT

The committee has identified the basic attributes of a systemwide modeling analysis that would support a useful mid-course assessment of CERP outcomes, given new information and changing conditions. The assessment should look into the future, beyond completion of the CERP, when sea level and temperatures will be higher, rainfall may be more variable, saltwater will have intruded farther into coastal aquifers, and freshwater demand may be greater. Ideally, the restored system should be resilient to stresses expected to arise by 2050 and for the remainder of the 21st century. The mid-course assessment should leverage new and updated hydrologic and ecological models and improved climate-model and sea-level projections that reflect knowledge and data accumulated since the CERP was authorized nearly 20 years ago. The update should also account for new information about CERP project feasibility, operations, and performance. The assessment should be integrative, including major CERP and non-CERP projects and treating them in a coupled, rather than independent, fashion. This integration is essential to examine effects of hydrologic and water quality interconnectedness in shaping Everglades-wide responses in the context of 21st-century conditions.

An assessment that incorporates a few of these basic attributes is in the planning phase. The RECOVER Five-Year Plan includes analysis to support the revision of the existing CERP Interim Goals and Interim Targets (RECOVER, 2005). These will be model-derived quantifications of expected ecological changes or other water-related services (e.g., water supply), based on projections of CERP implementation at future time intervals. RECOVER has gained approval for two sets of hydrologic and ecological modeling runs. The first set will include those projects that, according to the 2016 Integrated Delivery Schedule (see Chapter 3), have been authorized and are scheduled for completion by 2026. The second set will include all authorized projects scheduled for completion by 2030. All totaled, 30 projects are tentatively planned for inclusion in these model runs (Table 6-2). These two runs will be compared against a "current conditions base." The RECOVER analysis takes advantage of an impressive array of existing models. The South Florida Water Management Model (SFWMM) and the Regional Simulation Model (RSM) are among the models available to simulate water levels and flows. These hydrologic models provide the input data for ecological models capable of simulating vegetation in estuaries, Lake Okeechobee, and Everglades wetlands; aquatic fauna populations; and wading-bird nesting patterns.

This modeling effort represents a step toward a systemwide assessment and should inform an important and overdue update for the Interim Goals and Interim Targets. However, it falls short of what is needed for the mid-course assessment in several key respects. Hydrologic models for the RECOVER update will be forced by rainfall, evapotranspiration, and boundary conditions for a 1965-2005 period of record (W. Wilcox, SFWMD, personal communication, 2018). Although an important part of the analysis, use of the historical period of record alone to drive the hydrologic models is insufficient because there is no accounting for interactions among multidecadal and interannual climate variability, or changes in sea level and climate, that will manifest through the operational lifetimes of the restoration infrastructure. A suitably comprehensive assessment should, in addition to considering historic climate, consider a range of sea-level rise, temperature, and precipitation scenarios for the near term (e.g., 2020-2050), as major projects are completed, and the far term (e.g., 2050-2080),

TABLE 6-2 Potential Projects for Recover Five-Year Plan Modeling

IGIT Scenario Run	Projects Included	
Projects to be completed by 2026	Herbert Hoover Dike	Restoration Strategies
	Tamiami Trail Next Steps	Kissimmee River Restoration
	C-111 South Dade	Site 1 Impoundment – Phase 1
	C-44 Reservoir	C-44 STA
	C23/24 Reservoir North	C-25 Reservoir
	C-43 West Basin Storage	Broward Co WPA – Northern Mitigation Area
	Broward Co WPA – C-11 Impoundment	Picayune Strand Restoration
	Biscayne Bay Coastal Wetlands – Phase 1	C-111 Spreader Canal Western Project
	Old Tamiami Trail Modifications	L67A Structures + Gap in L-67C Levee (CEPP)
	Increase S-356 (CEPP)	L-29 Gated Spillway (CEPP)
	Increase S-333 (CEPP)	
Projects to be completed by 2030	All projects in 2026 run plus:	
	C23/C24 Reservoir South	C-25 STA
	Broward Co WPA – C-9 Impoundment	L-29 Levee Removal + L-67 Ext Backfill (CEPP)
	Broward Co WPA – 3A 3B Seepage Management	Construct L-67D Levee (CEPP)
	Remove L-67C + L-67 Ext (CEPP)	PPA North (CEPP)
	PPA New Water (CEPP)	

SOURCE: A. McLean, NPS, personal communication, 2018; D. Crawford, USACE, personal communication, 2018.

after these projects are operational and as the ecosystem responds to the restoration measures, sea-level rise, and climate change. This analysis could draw from significant efforts conducted outside of the CERP regarding climate-change impacts on South Florida's water resources, stormwater, and flood-management infrastructure (e.g., Dessalegne et al., 2016; Havens and Steinman, 2015; Irizarry et al., 2013; Nungesser et al., 2015; Obeysekera, 2013; Obeysekera et al., 2015; Park et al., 2017; Salas et al., 2018).

Another shortcoming of the RECOVER analysis lies in its restrictive focus on those CERP projects that are currently authorized. Consequently, it excludes assessment of potential benefits of major CERP projects that are now in the late stages of planning, such as the Lake Okeechobee Watershed Restoration Project and the Western Everglades Restoration Project as well as the recently authorized EAA Storage Reservoir Project. These projects will affect a large portion of the Everglades footprint. A comprehensive mid-course assessment should, at the outset, consider authorized projects and then be extended to consider projects in planning. It should also evaluate a range of possible Lake Okeechobee regulation schedules, beyond LORS 2008 and the prior WSE regulation schedules. Although the Lake Okeechobee regulation schedule is not slated for revision until 2022, this broader approach to a CERP mid-course assessment would shed light on how changed lake storage interacts with other projects and influences restoration outcomes.

Modeling for a mid-course, systemwide assessment should explore the near-term (2020-2050) and far-term (2050-2080) performance of the system under historic climate conditions and several future climate and sea-level rise scenarios for several CERP implementation scenarios. These CERP implementation scenarios could include

- Future without any CERP projects,
- Future with CERP projects as completed today,
- Future with all authorized CERP projects,
- Future with authorized CERP projects and CERP projects in planning, and
- Future with authorized and planned CERP projects plus potential alternative Lake Okeechobee regulation schedules.

A comparison among these climate and implementation scenarios would show the incremental benefits provided by CERP implementation and the sensitivity of these outcomes to climate change. Using several scenarios that encompass uncertainty about future climate, sea-level rise, and implementation enables exploration of what the future could hold for the CERP and provides a context for future planning and implementation, based on the current state of knowledge.

Future climate and sea level–rise scenarios could be defined from modeled projections that assume different representative concentration pathways (RCPs). The RCPs are four scenarios for greenhouse gas concentration made on the basis of expectations for 21st-century population change, income growth, technological improvements, and climate policies. The CERP mid-course assessment should consider future climate and sea level under two or more of these scenarios. For example, these could include RCP4.5, a relatively optimistic scenario, where deployment of policies and technologies mitigate emissions and stabilize radiative forcing, and RCP8.5, a high-end emissions scenario where emissions steadily increase over time.

Future climate assuming these emissions scenarios can be forecast with outputs from General Circulation Models (GCMs). The coarse-resolution GCM outputs are typically downscaled to higher spatial resolution using empirical-statistical methods, or they are used as boundary conditions in regional climate models that, in turn, yield outputs at high spatial resolution (DiNapoli and Misra,

2012; Misra et al., 2012b; Salathe et al., 2007; Selman et al., 2013; Wood et al., 2004). Downscaled climate projections from different GCMs never completely agree, leading to a key source of uncertainty in GCM-based climate forecasts. This uncertainty raises questions about how climate projections should be used in coupled hydrologic and ecological simulations. One answer is to weight climate model projections based on how well they match historical observations of pertinent climate variables, such as rainfall and temperature using a reliability ensemble averaging approach (Giorgio and Mearns, 2002) or Bayesian weighting approach (Tebaldi et al., 2005). An alternative to this performance-based approach—the so-called envelope approach—involves selecting a subset of GCM-based projections that cover the range of possible rainfall and temperature futures represented collectively in a large pool of climate models (Cannon et al., 2015; Immerzeel et al., 2013; Warszawksi et al., 2013). Obeysekera (2013) compared downscaled-GCM output to 20th-century observations for South Florida and showed that no single GCM simulated rainfall and temperature accurately enough for water-resource planning purposes. This realization led Obeysekera et al. (2015) to employ a simplified variant of the envelope approach to prescribe two mid–21st-century rainfall scenarios as a uniform ± 10% change around the historical rainfall time series. These rainfall scenarios, together with specifications of a 1.5°C temperature increase and 1.5 ft rise in sea level over historic levels, were used in the SFWMM to simulate the responses of Everglades water levels and flows to climate change without any restoration projects in place.

Approaches for using different climate scenarios that are conditional on the spread in climate-model projections are available, have been tested to a limited extent in South Florida, and should be adopted for the CERP mid-course assessment. The spread in climate forecasts for South Florida may decrease with continued research and improvements in the representation of modeled processes, but considerable uncertainty about future climate will likely persist. Because the spread (uncertainties) in climate forecasts will propagate into hydrologic predictions, results of a coupled climate-hydrologic analysis will yield a spectrum of potential hydrologic futures. This spectrum could shift or even broaden when interactions with alternative projections for sea-level rise are incorporated into the analyses. Nevertheless, the complicating effects of climate change and sea-level rise should not be ignored. Rather they should be characterized and brought to the fore using decision making under deep uncertainty (DMDU) planning processes such as robust decision making (RDM) (Groves and Lempert., 2007; Lempert et al., 2006). These approaches can be used to identify projects that do or do not meet management goals under a large number of climate change and sea level–rise scenarios. Such decision support tools can help inform the design of restoration infrastructure that will be durable through the 21st century and sufficiently flexible to perform under conditions that may be very different from those today.

Evaluation of the hydrologic and ecological outcomes achieved by each climate and implementation scenario would demonstrate the effect of projects already constructed and outcomes that might be achieved under future conditions following completion of major increments of the CERP. It would also illuminate ecosystem goals that are unlikely to be met as implementation of the current plan proceeds, providing a foundation for potential adjustment of CERP planning and implementation to achieve desirable outcomes in the face of climate change and to enable targeting of future investments where they can make a difference to the system for decades to come.

Outside of the CERP, the SFWMD is developing impressive expertise in decision making under deep uncertainty[2] that provides excellent frameworks for planning robust projects that deliver desired benefits across a wide range of possible future conditions. Following the mid-course assessment, the CERP agencies could use this expertise to inform future decision making.

SCIENCE TO ADVANCE THE CERP

Scientific understanding is fundamental to all aspects of Everglades restoration. It was crucial to the original development of the CERP and key advances made in the early years, such as the development of conceptual ecological models, were innovative and groundbreaking. Scientific knowledge is still advancing on many fronts, and new tools and approaches are being applied to yield insight on system dynamics and to support planning. However, there is an ongoing need for research and tool development to understand system change and how restoration affects it. This section presents several areas of research and tool development that would be useful components of a forward-looking science program to better support the CERP. These should be considered examples of the types of important issues that should be addressed. The committee also describes a programmatic approach to better support forward-looking research and development that is essential to the long-term success of CERP.

Research Needs to Support the Future Success of the CERP

Science support for the CERP, with its large and long-term infrastructure investments, requires attention to future stressors, their potential impacts on the South Florida ecosystem, and the implications for restoration. This section high-

---

[2] See www.deepuncertainty.org.

lights examples of important research and tool development needs to support the future success of the CERP.

Improved Modeling of Coastal Boundaries in Regional Models

Given the vulnerability of South Florida to sea-level rise, there is a critical need to advance tools and research to better characterize and quantify its effects. The current regional hydrologic model used by the SFWMD (RSM) has been a valuable tool to estimate historical discharges through the Everglades and to project the hydrologic response of planned or potential water management and restoration strategies. RSM has also been useful in making preliminary assessments of potential seasonal and spatial changes in water stage and flows under different climate change scenarios (Obeysekera et al., 2015). One drawback of this tool is the lack of a coupled connection with the coastal system, specifically the ability to simulate salinity transport and variable density flow. Improving the capabilities of hydrologic modeling tools to capture the coupled interaction with changing coastal boundary conditions, including salinity transport, would support assessment of the impacts of sea-level rise and storm surges on the Everglades ecosystem for current as well as future conditions. Improving the model to address changing coastal boundary conditions would also improve the capacity to evaluate the benefits provided to near coastal areas by restoration alternatives. Such enhanced modeling capabilities can be used to examine Everglades restoration options to improve freshwater flows while depicting the interface with the marine environment, enhancing the reliability of future planning and project evaluation. Additional data along the coastal boundary would also be needed to calibrate the modeling tools used to simulate the impacts of the changing coastal boundary conditions.

Understanding Peat Collapse

An important uncertainty involves the response of coastal vegetation and peat to sea-level rise. Field research is under way on wetland response to seawater inundation through the Florida Everglades Long-term Ecological Research (LTER) program, which should provide a better understanding of the response of freshwater wetlands and peat soils to inputs of marine water. These experiments could supply important quantitative and process-based information on the phenomenon of peat collapse. The committee encourages continued landscape-scale research on the dynamics of coastal wetland ecosystems with seawater inundation. For example, can the landward migration of mangroves keep pace with sea-level rise? What is the mechanism(s) and timescale of peat collapse, and how and at what rate do wetland ecosystems recover from this disturbance? Will the disturbance of seawater inundation alter the dynamics of phosphorus in these wetland ecosystems? This information will be critical to understanding the potential for freshwater flows to mitigate peat collapse.

Emerging remote sensing methods should prove useful in monitoring the landscape-scale dynamics of wetland response to saltwater inundation. An example of recent research along these lines is NASA Goddard's Lidar, Hyperspectral, and Thermal (G-LiHT) airborne imager (D. Lagomasino, University of Maryland, personal communication, 2018), which has been used to map the composition, structure, and function of terrestrial ecosystems, such as the mangrove forests along the coastal Everglades (Figure 6-6). G-LiHT data products and higher-level change maps are available through the G-LiHT Data Center.[3] Data from tools such as this will be essential to tracking lateral shifts in habitats with sea-level rise and freshwater flow restoration.

An important output of additional research and monitoring of vegetation and soil response to sea-level rise would be the coupling of a wetland landscape model with a revised hydrologic model that integrates the dynamics of sea-level rise in hydrologic simulations. A landscape submodel capable of depicting the accumulation and loss of peat and changes in land surface elevation in response to changes in freshwater flows as well as seawater inundation would be a valuable tool for CERP agencies.

Risk Assessment for Invasive Species

The identification and management of invasive species in South Florida will continue to be challenging. Recent advances in EDRR (early detection and rapid response) (e.g., Crall et al., 2012; Thomas et al., 2017) demonstrate that this is an active area of research, and CERP agencies and their partners should remain at the forefront of this work. For example, one increasingly used method to evaluate the likelihood that a nonnative fish species will become established is the Fish Invasiveness Screening Kit (FISK) (Hill et al., 2017; Lawson et al., 2015). This kit, like other assessment tools, is a systematic arrangement and evaluation of information about species that considers aspects of their biology, likelihood of spreading in the ecosystem, and other factors to assess the risk that they pose to a particular ecosystem. Retrospectively applying the tool to 95 nonnative species that had been introduced into U.S. public waters, Lawson et al. (2015) found that it correctly classified 76 percent of invasive species and 88 percent of noninvasive species. They concluded that managers could use the tool to identify nonnative species likely to become invasive. Tools like this

---

[3] See https://gliht.gsfc.nasa.gov.

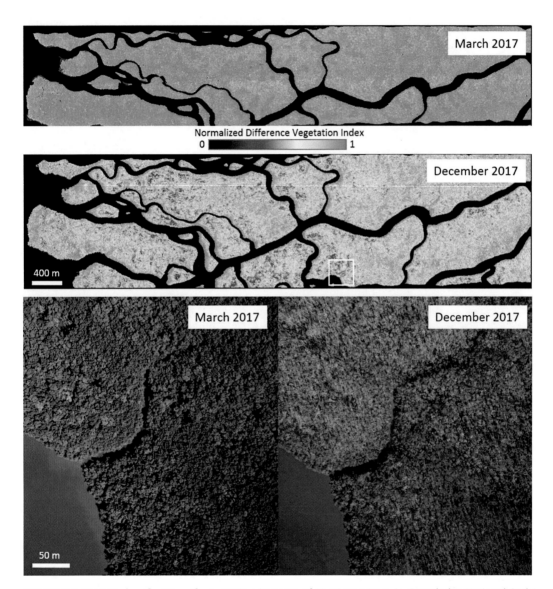

FIGURE 6-6 G-LiHT classification of vegetation impacts of Hurricane Irma in Everglades National Park/Shark River.

NOTES: Top: Normalized Difference Vegetation Index (NDVI) pre-Irma conditions (March 2017). Middle: NDVI post-Irma conditions (December 2017). Of the 8 km² swath of mangrove forest shown here, 60 percent was found to be heavily to severely damaged due to the hurricane. Aerial photos of the highlighted quadrant are shown for comparison.

SOURCE: D. Lagomasino, University of Maryland, personal communication, 2018.

one could also guide research on established exotic species by evaluating the potential value of different kinds of information in a risk management context.

This approach is starting to be applied in the Everglades. For example, a screening tool has been developed to assess the need to initiate a rapid response, as well as the resources and knowledge available to support that response (see Chapter 3).

Delivering Decision-Relevant Scientific Advances

An array of providers supports the technical effort underlying CERP planning and implementation (e.g., SFWMD, National Park Service, U.S. Geological Survey, National Science Foundation, Sea Grant, other agencies and nongovernmental organizations). Work focused on the Everglades is largely conducted by scientists and engineers in government agencies and universities based in South Florida, although work conducted outside the system, such as on climate change prediction, is tapped when applicable.

The RECOVER program is often seen as the centerpiece of the Everglades science endeavor. In the early years, as agencies started to implement the CERP, it served as a focal point of innovation with leaders in science who could break new ground and link scientific understanding to emerging management needs. The Programmatic Regulations task RECOVER with adaptive assessment and monitoring. The group works to "organize and apply scientific and technical information in ways that are most effective in supporting the objectives of CERP."[4] Prior to project implementation, this work provided flexibility for the development of tools and approaches such as conceptual models. Now that project implementation is in full swing, the emphasis has shifted toward supporting project planning and evaluation of benefits. RECOVER has produced an impressive array of reports to support restoration, and the RECOVER Five-Year Plan (RECOVER, 2016), to some extent, recognizes the need to reframe some of the work of the early 2000s to address the current and future needs of the program. Managing the massive endeavor of monitoring the restoration of the Everglades, supporting the simultaneous planning of multiple projects, and delivering required reports inevitably means that project management has become as important a function for RECOVER leaders as scientific vision.

Although ongoing monitoring, evaluation, synthesis, and reporting are essential tasks for tracking the progress of CERP implementation, there is a need to focus research and development activities on the implementation issues to come. The accelerated 3-year project planning process has moved projects forward quickly, but it provides minimal time to develop new tools and approaches,

---

[4] See http://www.saj.usace.army.mil/Missions/Environmental/Ecosystem-Restoration/RECOVER/.

digest the insights provided by new research, or ensure that Everglades science stays on the cutting edge. For science to successfully underpin Everglades restoration for decades to come, scientists working on Everglades restoration must be able to develop concepts and tools that future projects, as well as programmatic assessment, will need. The South Florida Ecosystem Restoration Task Force chartered the South Florida Ecosystem Restoration Science Coordination Group in 2003 to develop a report that "tracks and coordinates programmatic level science and other research, identifies programmatic level priority science needs and gaps, and facilitates management decisions" and to provide scientific support to the Task Force. The most recent Plan for Coordinating Science was updated in 2010 (SFERTF, 2010). It not only embraces the need for monitoring, evaluation, and assessment, but also explicitly acknowledges research and modeling as key elements of restoration science. Recent meetings of the Science Coordination Group have been focused on specific issues such as the revision of the conceptual ecological models and a coordinated science response to Hurricane Irma, which are worthwhile but insufficient to frame long-term research needs.

Currently, given all of the ongoing activities that comprise the Everglades science enterprise, there is no obvious method to formulate, let alone realize, this vision for science. The complexity and scale of the system make it challenging to incorporate emerging science into the restoration program and to ensure the availability of cutting-edge and usable science for implementing agencies and resource managers. This issue could be addressed through establishment of a specific science program to support the future implementation of the CERP. Meeting this challenge requires designation of the responsibility for developing and making available the body of knowledge necessary to support restoration activities. This program would not replace the work of individuals in RECOVER and the agencies who currently conduct the investigations, analyses, and model development that support effective restoration planning and implementation.

The scientific community within the Everglades, including university researchers and nongovernmental and private-sector experts, is already informally connected through the highly successful biennial Greater Everglades Ecosystem Restoration conference. This event provides an opportunity for "bottom-up" identification of emerging scientific issues and has proven to be such a successful vehicle for supporting collaboration, communication, and dissemination of scientific developments that it has been adopted in the California Bay-Delta[5] and in coastal Louisiana.[6] An Everglades "science program" could convene additional exchanges to pursue promising ideas that fall outside the purview of RECOVER or specific agencies.

---

[5] See http://scienceconf2018.deltacouncil.ca.gov/.
[6] See http://www.stateofthecoast.org/.

Although the Science Coordination Group facilitates science coordination related to ongoing activities, a designated science program could be a central point for setting a CERP-relevant science agenda that looks beyond the needs of the moment. The focus should be on specific research investigations to fill important gaps or drive innovative approaches to support restoration and could be pursued through a competitive process and leveraging of existing agency scientific investments. Program funds could be used, for example, to enhance direct collaboration between university and agency scientists or to pilot promising new concepts or approaches. The vision for the effort should not be difficult to develop. Everglades restoration agencies have a number of broadly knowledgeable, forward-thinking scientists who provide what CERP-level leadership they can, given their current positions and responsibilities. However, none of these is designated or empowered to take the reins of leading Everglades science forward for the next few decades.

Maintaining a focus on the importance of science for the future Everglades could be accomplished by establishing a new independent position of Lead Scientist. This individual would be responsible for high-level communication, delivery of technical products that respond to changing needs, and promotion of cutting-edge science in Everglades restoration—thereby freeing up time for individuals working in RECOVER or the agencies to deliver information to support day-to-day restoration and reporting activities. Again, there are several very capable, even visionary, senior scientists who provide leadership within their agencies, assume multiple roles within the CERP, and provide invaluable insight for Everglades restoration as a whole. However, there is no central leader to support Everglades restoration fully focused on a vision for science, its continued development, and application across agencies.

The Delta Stewardship Council in California provides a model for the Lead Scientist role. There, the Lead Scientist is appointed for a 2- to 3-year term, is often an individual on leave from a university position, and is responsible for leading, overseeing, and guiding the Council's Science Program. Lead Scientists for the Delta Stewardship Council are credentialed Ph.D.'s who have themselves made substantial contributions, and are skilled in communicating to policy makers and respected by their peers. Limiting the term of the position would not only prevent burnout in such a central, high-visibility role, but also allow for fresh insights to guide the Everglades scientific community.

Establishing a science program and designating a single leader for science, who can step aside from their university or agency setting and focus on the available and needed science, could empower and invigorate the broader scientific community. The Lead Scientist for the Delta Stewardship Council serves as a spokesperson for the broad science community, which is especially important when unpopular or politically charged scientific issues need to be communicated. Because of this role and administrative structure, the Delta Stewardship Council's Science Program is viewed by all stakeholders as the "honest brokers" of science. Such a program for the Everglades could also embrace the following activities:

- Guiding the development and refinement of effective integrative modeling tools,
- Charting the transition to new technologies and tools as they become available and tested for readiness,
- Prioritizing the array of issues and uncertainties that could be researched to support restoration,
- Identifying needs for advances in synthesis, modeling, and analytics to improve the capacities and responsiveness of the adaptive management program, and
- Providing a single voice for communication of science at the highest level. For example, the Delta Stewardship Council Lead Scientist serves on the Delta Plan Interagency Implementation Committee and regularly testifies to legislative and policy bodies.

Such an approach would return Everglades restoration to the forefront of science-informed restoration (see NASEM, 2016). A recent workshop summary (USGS and Delta Stewardship Council, 2018) identified clear and effective science leadership and relationship building as critical to the success of any restoration science enterprise. For such a position to be successful in the Everglades will require backing and cooperation across agencies, and in turn the Lead Scientist must appreciate the differing roles and responsibilities of the agencies involved.

The committee recognizes the barriers to funding such a science program, given the CERP's project-specific funding approach. However, relatively modest funds from the myriad of partners with spending flexibility and a direct interest in the science could be pooled to support a Lead Scientist and modest staff. Funding for research and development already exists but it not well coordinated, and pooling dollars could lead to efficiencies and lower the need for future funding requests.

CONCLUSIONS AND RECOMMENDATIONS

The Everglades of 2050 and beyond will differ from what was originally envisioned when the CERP was developed. The original CERP plan was formulated based on a pre-drainage vision of the historical Everglades and the assumption that specific rainfall and temperature time series observed in the past captured the full range of variability expected throughout the 21st century. There is now ample evidence that the South Florida climate is changing. There is general consensus that temperatures will increase over time, although considerable uncertainty about future rainfall patterns remains. There is also compelling recent evidence that sea-level rise is accelerating. These changes will have profound impacts on the South Florida ecosystem and the related challenges of providing flood protection and meeting future water and recreational demands.

CERP agencies should conduct a mid-course assessment that rigorously considers the future of the South Florida ecosystem. New information about climate variability, climate change, and sea-level rise in South Florida continues to emerge, and many of these changes will impact the capacity for the CERP to meet its goals. Although the SFWMD has begun to conduct these types of analyses for planning and management projects outside of the restoration, CERP agencies do not adequately account for these changes when planning projects, and they have not systematically analyzed these threats in the context of the CERP. Restoration is likely to create important benefits that increase the resilience of the ecosystem in the face of climate change, but these benefits have not been adequately studied or quantified. A systemwide, program-level analysis should assess the resilience and robustness of the CERP to the changing conditions that will drive the Everglades of the future. A mid-course assessment should include systemwide modeling of interactions among both authorized and planned projects under scenarios of future possible climate and sea level–rise conditions. This assessment is essential to communicate the benefits of the CERP to stakeholders, guide project sequencing and investment decisions, and manage the restoration under changing conditions. Now that several major project planning efforts are nearing completion and the vision for CERP storage is largely developed, which will require decades to construct at current funding levels, the time is right for a mid-course assessment.

A science program focused on understanding impacts of current and future stressors on the South Florida ecosystem is needed to ensure that CERP agencies have the latest scientific information and tools to successfully plan and implement the restoration program. This report has highlighted the ongoing research advances and science that are needed to address issues of vital importance for the long-term success of restoration investments, such as understanding peat collapse, saltwater intrusion, and the management of invasive species. Ensuring that investigative research and advances in tools and understanding are useful in a policy context requires a programmatic approach directly linked to the CERP effort, which may be best championed by an independent Everglades Lead Scientist empowered to coordinate and promote needed scientific advances.

References

Aldridge, F. J., E. J. Phlips, and C. L. Schelske. 1995. The use of nutrient enrichment bioassays to test for spatial and temporal distribution of limiting factors affecting phytoplankton dynamics in Lake Okeechobee, Florida. *Archiv für Hydrobiologie, Advances in Limnology* 4: 177-190.

Ali, A. 2009. Nonlinear multivariate rainfall-stage model for large wetland systems. *Journal of Hydrology* 374(3-4): 338-350.

Alongi, D. M. 2008. Mangrove forests: Resilience, protection from tsunamis, and responses to global climate change. *Estuarine, Coastal and Shelf Science* 76(1): 1-13.

Audubon of Florida. 2010. Site 1 Impoundment on the Fran Reich Preserve: Restoring the Health of the Arthur R. Marshall Loxahachee NWR. Information Sheet, October. http://fl.audubon.org/sites/g/files/amh666/f/revised-site-1-project-groundbreaking-audubon-factsheet.pdf.

Aumen, N. G., and R. E. Wetzel. 1995. Ecological studies on the littoral and pelagic systems of Lake Okeechobee, Florida (USA). *Archiv für Hydrobiologie, Advances in Limnology* 45: 356.

Barry, M. J., M. S. Bonness, and C. Heiden. 2017. Attachment F: Year 9 Post-Restoration Vegetation Monitoring of Select Eastern Picayune & Control Transects—Picayune Strand Restoration Project. PSRP Vegetation Monitoring 2016. 2018 South Florida Environmental Report—Volume III, Appendix 2-1. November 30, 2017.

Batzer, D. P., and R. R. Shartiz. (eds.). 2006. Ecology of Freshwater and Estuarine Wetlands. Berkeley: University of California Press. doi: 10.1525/california/9780520247772.003.0001.

Bennetts, R. E., and W. Kitchens. 1997. Demography and Movements of Snail kites in Florida. Technical Report 56, Florida Cooperative Fish and Wildlife Research Unit, Gainesville, FL.

Blake, N. 1980. Land into Water—Water into Land: A History of Water Management in Florida. Tallahassee: University Presses of Florida.

Brandt, L. A., J. Beauchamp, J. A. Browder, M. Cherkiss, A. Clarke, R. F. Doren, P. Frederick, E. Gaiser, D. Gawlik, L. Glenn, E. Hardy, A. L. Haynes, A. Huebner, K. Hart, C. Kelble, S. Kelly, J. Kline, K. Kotun, G. Liehr, J. Lorenz, C. Madden, F. J. Mazzotti, L. Rodgers, A. Rodusky, D. Rudnick, B. Sharfstein, R. Sobszak, J. Trexler, and A. Volety. 2014. System-wide Indicators for Everglades Restoration. Unpublished Technical Report. 111 pp. https://www.evergladesrestoration.gov/content/documents/system_wide_ecological_indicators/2014_system_wide_ecological_indicators.pdf.

Browder, J., R. Alleman, S. Markley, P. Ortner, and P. Pitts. 2005. Biscayne Bay conceptual ecological model. *Wetlands* 25: 854-869.

Camp, E. V., W. E. Pine, K. Havens, A. S. Kane, C. J. Walters, T. Irani, A. B. Lindsay, and J. G. Morris, Jr. 2015. Collapse of a historic oyster fishery: Diagnosing causes and identifying paths toward increased resilience. *Ecology and Society* 20: 45-57.

Cannon, A. J., S. R. Sobie, and T. Q. Murdock. 2015. Bias correction of GCM precipitation by quantile mapping: How well do methods preserve changes in quantiles and extremes? *Journal of Climate* 28:6938-6959.

Carter, L. M., J. W. Jones, L. Berry, V. Burkett, J. F. Murley, J. Obeysekera, P. J. Schramm, and D. Wear. 2014. Chapter 17 in Southeast and the Caribbean. Climate Change Impacts in the United States: The Third National Climate Assessment, J. M. Melillo, Terese (T. C.) Richmond, and G. W. Yohe, Eds. U.S. Global Change Research Program. doi: 10.7930/J0N-P22CB.

Castañeda-Moya, E., R. R. Twilley, V. H. Rivera-Monroy, K. Zhang, S. E. Davis, and M. Ross. 2010. Spatial patterns of sediment deposition in mangrove forests of the Florida Coastal Everglades after the passage of Hurricane Wilma. *Estuaries and Coasts* 33: 45-58.

Castañeda-Moya, E., R. R. Twilley, V. H. Rivera-Monroy, B. D. Marx, C. Coronado-Molina, and S. M. L. Ewe. 2011. Patterns of root dynamics in mangrove forests along environmental gradients in the Florida Coastal Everglades, USA. *Ecosystems* 14(7): 1178-1195.

Castaneda-Moya, E., R. R. Twilley, and V. H. Rivera-Monroy. 2013. Allocation of biomass and net primary productivity of mangrove forests along environmental gradients in the Florida Coastal Everglades, USA. *Forest Ecology and Management* 307: 226-241. doi: 10.1016/j.foreco.2013.07.011.

CCSP (U.S. Climate Change Science Program). 2009. Coastal Sensitivity to Sea-Level Rise: A Focus on the Mid-Atlantic Region. A report by the U.S. Climate Change Science Program and the Subcommittee on Global Change Research. J. G. Titus (Coordinating Lead Author), K. E. Anderson, D. R. Cahoon, D. B. Gesch, S. K. Gill, B. T. Gutierrez, E. R. Thieler, and S. J. Williams (Lead Authors). U.S. Environmental Protection Agency. Washington, DC, 320 pp.

Chambers, L. G., K. R. Reddy, and T. Z. Osborne. 2011. Short-term response of carbon cycling to salinity pulses in a freshwater wetland. *Soil Science Society of America Journal* 75: 2000-2007.

Chambers, L. G., T. Z. Osborne, and K. R. Reddy. 2013. Effect of salinity-altering pulsing events on soil organic carbon loss along an intertidal wetland gradient: A laboratory experiment. *Biogeochemistry* 115: 363-383.

Chambers, L. G., S. E. Davis, T. G. Troxler, J. N. Boyer, A. Downey-Wall, and L. J. Scinto. 2014. Biogeochemical effects of simulated sea level rise on carbon loss in an Everglades mangrove peat soil. *Biogeochemistry* 726(1): 195-211.

Chambers, L. G., S. E. Davis, and T. G. Troxler. 2015. Sea level rise in the Everglades: Plant-soil microbial feedbacks in response to changing physical conditions. Chapter 5 in Microbiology of the Everglades Ecosystem. Boca Raton, FL: CRC Press.

Charkhian, B., D. Hazelton, S. Blair, J. Mahoney, J. Possley, G. Garis, L. Baldwin, B. Gu, L. Waugh, and R. Pfeuffer. 2014. Appendix 2-3: Annual Permit Report for the Biscayne Bay Coastal Wetlands—L31 East Culverts and Deering Estate Flow-way. 2015 South Florida Environmental Report. South Florida Water Management District, West Palm Beach, FL. http://apps.sfwmd.gov/sfwmd/SFER/2014_SFER/v3/appendices/v3_app2-3.pdf.

Charkhian, B., D. Hazelton, S. Blair, J. Mahoney, J. Possley, L. Baldwin, B. Gu, and L. Waugh. 2015. Appendix 2-3: Annual Permit Report for the Biscayne Bay Coastal Wetlands—L31 East Culverts and Deering Estate Flow-way. 2015 South Florida Environmental Report. South Florida Water Management District, West Palm Beach, FL. http://apps.sfwmd.gov/sfwmd/SFER/2015_sfer_final/v3/appendices/v3_app2-3.pdf.

Charkhian, B., D. Hazelton, C. Grossenbacher, J. Mahoney, J. Possley, L. Baldwin, B. Gu, and L. Waugh. 2016. Appendix 2-3: Annual Permit Report for the Biscayne Bay Coastal Wetlands Project. 2016 South Florida Environmental Report. South Florida Water Management District, West Palm Beach, FL. http://apps.sfwmd.gov/sfwmd/SFER/2016_sfer_final/v3/appendices/v3_app2-3.pdf.

Charkhian, B., D. Hazelton, J. Mahoney, J. Possley, L. Baldwin, B. Gu, and P. Kamthe. 2017. Appendix 2-3: Annual Permit Report for the Biscayne Bay Coastal Wetlands Project. 2017 South Florida Environmental Report. Volume III. South Florida Water Management District, West Palm Beach, FL. http://apps.sfwmd.gov/sfwmd/SFER/2017_sfer_final/v3/appendices/v3_app2-3.pdf.

Charkhian, B. 2017. Biscayne Bay Coastal Wetlands Phase I: Restoration Benefits Observed from the Biscayne Bay Coastal Wetlands Project. 2017 GEER (Greater Everglades Ecosystem Restoration), Coral Springs, FL. April 20, 2017.

Charkhian, B., L. Baldwin, B. Gu, D. Hazelton, M. Hodapp, and J. Possley. 2018. Appendix 2-3: Annual Permit Report for the Biscayne Bay Coastal Wetlands Project. 2018 South Florida Environmental Report. Volume III. South Florida Water Management District, West Palm Beach, FL. http://apps.sfwmd.gov/sfwmd/SFER/2018_sfer_final/v3/appendices/vv3_app2-3.pdf.

Chastant, J. E., M. L. Peterson, and D. E. Gawlick. 2017. Nesting substrate and water-level fluctuations influence wading bird nesting patterns in a large shallow eutrophic lake. *Hydrobiologia* 788: 371-383.

Chesapeake Bay Program. 2009. Development and Implementation of a Process for Establishing Chesapeake Bay Program's Monitoring Program Priorities and Objectives. http://www.chesapeake.org/pubs/237_2009.pdf.

Childers, D. L., J. N. Boyer, S. E. Davis, C. J. Madden, D. T. Rudnick, and F. H. Sklar. 2006. Relating precipitation and water management to nutrient concentrations in the oligotrophic "upside-down" estuaries of the Florida Everglades. *Limnology and Oceanography* 51: 602-616. doi: 10.4319/lo.2006.51.1_part_2.0602.

Choi, J., E. Cline, C. Coronado-Molina, M. Dickman, N. Dorn, C. Hansen, J. W. Harvey, M. Manna, S. Newman, B. H. Rosen, C. J. Saunders, F. H. Sklar, L. Soderqvist, E. Tate-Boldt, J. Trexler, and C. Zweig. 2017. The Decomp Physical Model (DPM) Science Plan: The Effects of Year-Round Sheetflow and Canal Backfilling. U.S. Army Corps—Jacksonville. https://erdc-library.erdc.dren.mil/xmlui/bitstream/handle/11681/26826/Supp%20EA%20%-26%20FONSI%20%20WCA%20Decomparmentalization%20Project_App%20C.pdf?sequence=1&isAllowed=y.

Chuirazzi, K., W. Abtew, V. Ciuca, B. Gu, N. Iricanin, C. Mo, N. Niemeyer, and J. Starnes. 2018. Appendix 2-1: Annual Permit Report for the Picayune Strand Restoration Project. 2018 South Florida Environmental Report—Volume III. http://apps.sfwmd.gov/sfwmd/SFER/2018_sfer_final/v3/appendices/v3_app2-1.pdf.

Coen, L. D., M. W. Luckenbach, and D. L. Breitberg. 1999. The role of oyster reefs as essential fish habitat: A review of current knowledge and some new perspectives. *American Fisheries Society Symposium* 22: 438-454.

Conner, M. M., W. C. Saunders, N. Bouwes, and C. Jordan. 2015. Evaluating impacts using a BACI design, ratios, and a Bayesian approach with a focus on restoration. *Environmental Monitoring and Assessment* 188(10): 555.

Convertino, M., K. Baker, C. Lu, J. T. Vogel, K. McKay, and I. Linkov. 2013. Metric Selection for Ecosystem Restoration. Ecosystem Management and Restoration Research Program EMRRP-EBA-19. https://el.erdc.dren.mil/elpubs/pdf/eba19.pdf.

Conyngham, J. 2010. Guidance on Monitoring Ecosystem Restoration Projects. Slides from Environmental Benefits Analysis Seminar, January 12, 2010. https://cw-environment.erdc.dren.mil/webinars/10jan12-Monitoring.pdf.

Coops, H., N. Geilen, H. Verheij, R. Boeters, and G. van der Velde. 1996. Interactions between waves, bank erosion and emergent vegetation: An experimental study in a wave tank. *Aquatic Botany* 53: 187-198.

COPC (California Ocean Protection Council). 2018. State of California Sea-Level Rise Guidance: 2018 Update. California Ocean Protection Council: Sacramento, CA. http://www.opc.ca.gov/webmaster/ftp/pdf/agenda_items/20180314/Item3_Exhibit-A_OPC_SLR_Guidance-rd3.pdf.

CPRA (Coastal Protection and Restoration Authority). 2012. Louisiana's Comprehensive Master Plan for a Sustainable Coast. Louisiana's Comprehensive Master Plan for a Sustainable Coast. Available at http://coastal.la.gov/a-common-vision/2012-coastal-master-plan/.

CPRA. 2017. Louisiana's Comprehensive Master Plan for a Sustainable Coast. Available at http://coastal.la.gov/wp-content/uploads/2017/01/DRAFT-2017-Coastal-Master-Plan.pdf.

Crall, A. W., M. Renz, B. J. Panke, G. J., Newman, C. Chapin, J. Graham, and C. Bargeron. 2012. Developing cost-effective early detection networks for regional invasions. *Biological Invasions* 14(12): 2461-2469.

Cronk, J. K., and M. S. Fennessy. 2001. Wetland Plants: Biology and Ecology. Boca Raton, FL: CRC Press/Lewis Publishers.

Davis, S. M., and J. C. Ogden (eds). 1994. Everglades: The Ecosystem and Its Restoration. Delray Beach, FL: St. Lucie Press.

Davis, S. M., D. L. Childers, J. J. Lorenz, H. R. Wanless, and T. E. Hopkins. 2005. A conceptual model of ecological interactions in the mangrove estuaries of the Florida Everglades. *Wetlands* 25: 832-842.

Day, J. W., G. P. Shaffer, L. D. Britsch, D. J. Reed, S. R. Hawes, and D. Cahoon. 2000. Pattern and process of land loss in Mississippi Delta: A spatial temporal analysis of wetland habitat change. *Estuaries* 23: 425-438.

DeLaune, R. D., J. A. Nyman, and J. W. H. Patrick. 1994. Peat collapse, ponding and wetland loss in a rapidly submerging coastal marsh. *Journal of Coastal Research* 104: 1021-1030.

Dessalegne, T., J. Obeysekera, S. Nair, and J. Barnes. 2016. Assessment of CMIP5 Multi-Model Database to Evaluate Impacts on the Future Regional Water Resources of South Florida. World Environmental and Water Resources Congress.

Dessu, S. B., R. M. Price, T. G. Troxler, and J. S. Kominoski. 2018. Effects of sea-level rise and freshwater management on long-term water levels and water quality in the Florida Coastal Everglades. *Journal of Environmental Management* 211:164-176.

DiNapoli, S., and V. Misra, 2012: Reconstructing the 20th century high-resolution climate of the southeastern United States. *Journal of Geophysical Research* 117(D19113). doi: 10.1029/2012JD018303.

Doering, P. H. 1996. Temporal variability of water quality in the St. Lucie Estuary, South Florida. *Journal of the American Water Resources Association* 32: 1293-1306.

Doering, P. H., and R. H. Chamberlain. 1999. Water quality and source of freshwater discharge to the Caloosahatchee Estuary, Florida. *Journal of the American Water Resources Association* 35: 793-806.

DOI (Department of the Interior) and USACE (U.S. Army Corps of Engineers). 2005. Central and Southern Florida Project Comprehensive Everglades Restoration Plan: 2005 Report to Congress.

Doren, R. F., J. C. Trexler, A. D. Gottlieb, and M. C. Harwell. 2009. Ecological indicators for system-wide assessment of the Greater Everglades Ecosystem Restoration Program. *Ecological Indicators* 9(6): S2-S16.

Dunne, E. J., M. W. Clark, R. Corstanje, and K. R. Reddy. 2011. Legacy phosphorus in subtropical wetland soils: Influence of dairy, improved and unimproved pasture land use. *Ecological Engineering* 37: 1481-1491.

Enfield, D. B., A. M. Mestas-Nunez, and P. J. Trimble. 2001. The Atlantic multidecadal oscillation and its relation to rainfall and river flows in the continental U.S. *Geophysical Research Letters* 28(10): 2077-2080.

EPA (Environmental Protection Agency). 2010. United States Environmental Protection Agency Amended Determination. http://peer.org/docs/fl/9_22_10_EPA_Amended_Determination.pdf.

Falk, D. A., M. A. Palmer, and J. B. Zedler. 2006. Ecological theory and restoration ecology. In Foundations of Restoration Ecology. D. A. Falk, Palmer, M. A., and J. B. Zedler, Eds. Society for Ecological Restoration International. Washington, DC: Island Press.

FDEP (Florida Department of Environmental Protection). 2001. The State of Florida's total phosphorus total maximum daily load (TMDL) for Lake Okeechobee. Florida Department of Environmental Protection, Tallahassee, FL.

FDEP. 2010. Loxahatchee River National Wild and Scenic River Management Plan: Update 2010. Florida Department of Environmental Protection and South Florida Water Management District, 98 pp.

FDEP. 2015. Report on the Beneficial Use of Reclaimed Water, Stormwater, and Excess Surface Water (Senate Bill 536). Office of Water Policy. December 1, 2015. http://www.dep.state.fl.us/water/reuse/docs/sb536/SB536-Report.pdf.

Fisher, M. M., K. R. R. Reddy, and R. T. James. 2001. Long-term changes in the sediment chemistry of a large shallow subtropical lake. *Lake and Reservoir Management* 17: 217-232.

Flaig, E. G., and K. E. Havens. 1995. Historical trends in the Lake Okeechobee ecosystem I. Land use and nutrient loading. *Archiv für Hydrobiologie Monographische Beitrage* 107: 1-24.

Flaig, E. G., and K. R. Reddy. 1995. Fate of phosphorus in the Lake Okeechobee watershed, Florida, USA: Overview and recommendations. *Ecological Engineering* 5: 127-142. http://www.sciencedirect.com/science/article/pii/0925857495000216.

Fletcher, R., C. Poli, E. Robertson, B. Jeffery, S. Dudek, and B. Reichert. 2017. Snail Kite Demography. 2016 Annual Report. Prepared for the USACE and Florida FWC. USGS Florida Cooperative Fish and Wildlife Research Unit, University of Florida, Gainesville.

Foster, B. 2018. Presentation of the Project Delivery Team Meeting #12: Loxahatchee River Watershed Restoration Project: Preview of Evaluation of Alternative Plans. U.S. Army Corps of Engineers—Jacksonville District.

Fry, B., P. L. Mumford, D. D. Fox, G. L. Warren, K. E. Havens, and A. D. Steinman. 1999. Trophic position and individual feeding histories of fish from Lake Okeechobee, Florida. *Canadian Journal of Fisheries and Aquatic Sciences* 56: 590-600.

FWC (Florida Fish and Wildlife Conservation Commission). 2015. Oyster monitoring in the northern estuaries on the southeast coast of Florida. Florida Fish and Wildlife Conservation Commission, June 2015, 37 pp.

FWS (Fish and Wildlife Service). 2016. Biological Opinion for the Everglades Restoration Transition Plan—2016. July 22, 2016. https://www.fws.gov/verobeach/verobeach_old-dont_delete/images/biologicalopinion/20160722ERTPBO.pdf.

George, D. 2016 (March). Everglades Restoration Transition Plan. USACE Public Meeting Presentation. http://my.sfwmd.gov/webapps/publicMeetings/viewFile/8740.

Giorgi, F., and L. O. Mearns. 2002. Calculation of average, uncertainty range, and reliability of regional climate changes from AOGCM simulations via the "reliability ensemble averaging" (REA) method. *Journal of Climate* 15: 1141-1158.

Gonzales, H. 2018. Presentation to the SFER Task Force: Program and Project Update. U.S. Army Corps of Engineers—Jacksonville District.

Graham, W. D., M. J. Angelo, T. K. Frazer, P. C. Frederick, K. E. Havens, and K. R. Reddy. 2015. Options to Reduce High Volume Freshwater Flows to the St. Lucie and Caloosahatchee Estuaries and Move More Water from Lake Okeechobee to the Southern Everglades: An Independent Technical Review by the University of Florida Water Institute.

Grimshaw, H. J., K. E. Havens, B. Sharfstein, A. Steinman, D. Anson, T. East, R. P. Maki, A. Rodusky, and K. R. Jin. 2002. The effects of shading on morphometric and meristic characteristics of *Vallisneria americana* transplants from Lake Okeechobee, Florida. *Archiv fur Hydrobiologie* 155: 65-81.

Groves, D. G., and R. J. Lempert. 2007. A New Analytic Method for Finding Policy-relevant scenarios. RAND Corporation. Santa Monica, CA. https://www.rand.org/pubs/reprints/RP1244.html.

Grunwald, M. 2006. The Swamp: The Everglades, Florida, and the Politics of Paradise. New York: Simon and Schuster.

Harwell, M. C., and K. E. Havens. 2003. Experimental studies on the recovery potential of submerged aquatic vegetation after flooding and desiccation in a large subtropical lake. *Aquatic Botany* 77: 135-151.

Harwell, M. C., and B. Sharfstein. 2009. Submerged aquatic vegetation and bulrush in Lake Okeechobee as indicators of greater Everglades ecosystem restoration. *Ecological Indicators* 9(6): S46-S55.

Havens, K. E. 2002. Development and application of hydrologic restoration goals for a large subtropical lake. *Lake and Reservoir Management* 18: 285-292.

Havens, K. E. 2003. Submerged aquatic vegetation correlations with depth and light attenuating materials in a shallow subtropical lake. *Hydrobiologia* 493: 173-186.

Havens, K. E., and D. E. Gawlik. 2005. Lake Okeechobee conceptual ecosystem model. *Wetlands* 25: 908-925.

Havens, K. E., and A. D. Steinman. 2015. Ecological responses of a large shallow lake (Okeechobee, Florida) to climate change and potential future hydrologic regimes. *Environmental Management* 55(4): 763-775.

Havens, K. E., and W. W. Walker. 2002. Development of a total phosphorus concentration goal in the TMDL process for Lake Okeechobee, Florida (USA). *Lake and Reservoir Management* 18: 227-238.

Havens, K. E., L. A. Bull, G. L. Warren, T. L. Crisman, E. J. Phlips, and J. P. Smith. 1996. Food web structure in a subtropical lake ecosystem. *Oikos* 75: 20-32.

Havens, K. E., M. C. Harwell, M. A. Brady, B. Sharfstein, T. L. East, A. J. Rodusky, D. Anson, and R. P. Maki. 2002. Large-scale mapping and predictive modeling of submerged aquatic vegetation in a shallow eutrophic lake. *The Scientific World Journal* 2: 949-965.

Havens, K. E., B. Sharfstein, M. A. Brady, T. L. East, M. C. Harwell, R. P. Maki, and A. J. Rodusky. 2004. Recovery of submerged plants from high water stress in a large subtropical lake in Florida, USA. *Aquatic Botany* 78: 67-82.

Havens, K. E., K. R. Jin, N. Iricanin, and R. T. James. 2007. Phosphorus dynamics at multiple time scales in the pelagic zone of a large shallow lake in Florida, USA. *Hydrobiologia* 581: 25-42.

Havens, K. E., J. R. Beaver, D. A. Cassamatta, T. L. East, R. T. James, P. McCormick, E. J. Phlips, and A. J. Rodusky. 2011. Hurricane effects on the planktonic food web of a large subtropical lake. *Journal of Plankton Research* 33: 1081-1094.

Havens, K., H. Paerl, E. Phlips, M. Zhu, J. Beaver, and A. Srifa. 2016. Extreme weather events and climate variability provide a lens to how shallow lakes may respond to climate change. *Water* 8. doi: 10.3390/w8060229.

Herbert, E. R., P. Boon, A. J. Burgin, S. C. Neubauer, R. B. Franklin, M. Ardón, K. N. Hopfensperger, L. P. M. Lamers, and P. Gell. 2015. A global perspective on wetland salinization: Ecological consequences of a growing threat to freshwater wetlands. *Ecosphere* 6(10): 206. http://dx.doi.org/10.1890/ES14-00534.1.

Hiers, J. K., R. J. Mitchell, A. Barnett, J. R. Walters, M. Mack, B. Williams, and R. Sutter. 2012. The dynamic reference concept: Measuring restoration success in a rapidly changing no-analogue future. *Ecological Restoration* 30: 27-36.

Hiers, J. K., S. Jackson, R. Hobbs, E. Bernhardt, and L. Valentine. 2016. The precision problem in conservation and restoration. *Trends in Ecology and Evolution* 31: 820-830.

Hill, J. E., Q. M. Tuckett, S. Hardin, L. L. Lawson, Jr., K. M. Lawson, J. L. Ritch, and L. Partridge. 2017. Risk screen of freshwater tropical ornamental fishes for the conterminous United States. *Transactions of the American Fisheries Society* 146: 927-938.

Huang, W., S. Hagen, P. Bacopoulos, and D. Wang. 2015. Hydrodynamic modeling and analysis of sea level rise impacts on salinity for oyster growth in Apalachicola Bay, Florida. *Estuarine, Coastal and Shelf Science* 156: 7-18.

Hwang, S. J., K. E. Havens, and A. D. Steinman. 1999. Phosphorus kinetics of planktonic and benthic assemblages in a shallow subtropical lake. *Freshwater Biology* 40: 729-745.

Immerzeel, W. W., F. Pellicciotti, and M. F. P. Bierkens. 2013. Rising river flows throughout the twenty-first century in two Himalayan glacierized watersheds. *Nature Geoscience* 6(9): 742.

IPCC (Intergovernmental Panel on Climate Change). 2014: Climate Change 2014: Synthesis Report. Contribution of Working Groups I, II and III to the Fifth Assessment Report of the Intergovernmental Panel on Climate Change, R. K. Pachauri and L. A. Meyer, Eds. Geneva, Switzerland: IPCC, 151 pp.

IPRL (Invasive Plant Research Laboratory). 2017. Biological Control Program: Comprehensive Everglades Restoration Plan. FY 17 Combined Final and 4th Quarterly (1 Jul - 30 Sep 2017) Report. Ft. Lauderdale: USDA/ARS Invasive Plant Research Laboratory.

Irizarry, M. M., J. Obeysekera, J. Park, P. Trimble, J. Barnes, W. Said, and E. Gadzinski. 2013. Historical trends in Florida temperature and precipitation. *Journal of Hydrological Processes* 27(16): 2225-2382.

Jacoby, M. 2018. Presentation at the SFER Task Force Meeting: SFWMD Program and Project Update. South Florida Water Management District. https://evergladesrestoration.gov/content/tf/minutes/2018_meetings/072518/7_SFWMD%20Program%20Project%20Update%20July%202018%20TF%20mtg%20-%20Megan%20Jacoby.pdf.

James, R. T., and C. D. Pollman. 2011. Sediment and nutrient management solutions to improve the water quality of Lake Okeechobee. *Lake and Reservoir Management* 27: 28-40.

Ji, G., K. E. Havens, J. R. Beaver, and T. L. East. 2018. Recovery of plankton from hurricane impacts in a large shallow lake. *Freshwater Biology*. doi: 10.1111/fwb.13075.

Jin, K. R., and D. Sun. 2007. Sediment resuspension and hydrodynamics in Lake Okeechobee during the late summer. *Journal ASCE Engineering* 133: 899-910.

Jin, K. R., and Z. G. Ji. 2001. Calibration and verification of a spectral wind–wave model for Lake Okeechobee. *Ocean Engineering* 28(5): 571-584.

Jin, K. R., and Z. G. Ji. 2013. A long term calibration and verification of a submerged aquatic vegetation model for Lake Okeechobee. *Ecological Processes* 2(1): 23.

Johnson, S. G., M. S. Allen, and K. E. Havens. 2007. A review of littoral vegetation, fisheries and wildlife responses to hydrologic variation at Lake Okeechobee. *Wetlands* 27: 110-126.

Johnson, D. H. 2012. Monitoring That Matters. Design and Analysis of Long-Term Ecological Monitoring Studies. R. A. Gitzen, J. J. Millspaugh, A. B. Cooper, and D. S. Licht, Eds. Cambridge and New York: Cambridge University Press.

Julian, P., G. G. Payne, and S. K. Xue. 2013. Chapter 3A: Water Quality in the Everglades Protection Area. 2013 South Florida Environmental Report—Volume 1. South Florida Water Management District, West Palm Beach, FL. http://apps.sfwmd.gov/sfwmd/SFER/2013_SFER/v1/chapters/v1_ch3a.pdf.

Julian, P., G. G. Payne, and S. K. Xue. 2014. Chapter 3A: Water Quality in the Everglades Protection Area. 2014 South Florida Environmental Report—Volume 1. South Florida Water Management District, West Palm Beach, FL. http://apps.sfwmd.gov/sfwmd/SFER/2014_SFER/v1/chapters/v1_ch3a.pdf.

Julian, P., G. G. Payne, and S. K. Xue. 2015. Chapter 3A: Water Quality in the Everglades Protection Area. 2015 South Florida Environmental Report—Volume 1. South Florida Water Management District, West Palm Beach, FL. http://apps.sfwmd.gov/sfwmd/SFER/2015_sfer_final/v1/chapters/v1_ch3a.pdf.

Julian, P., G. G. Payne, and S. K. Xue. 2016. Chapter 3A: Water Quality in the Everglades Protection Area. 2016 South Florida Environmental Report—Volume 1. South Florida Water Management District, West Palm Beach, FL. http://apps.sfwmd.gov/sfwmd/SFER/2016_sfer_final/v1/chapters/v1_ch3a.pdf.

Julian, P., A. Freitag, G. G. Payne, S. K. Xue, and K. McClure. 2018. Chapter 3A: Water Quality in the Everglades Protection Area. 2016 South Florida Environmental Report—Volume 1. South Florida Water Management District, West Palm Beach, FL. http://apps.sfwmd.gov/sfwmd/SFER/2018_sfer_final/v1/chapters/v1_ch3a.pdf.

Kearney, K., M. Butler, R. Glazer, C. Kelble, J. Serafy, and E. Stabenau. 2015. Quantifying Florida Bay habitat suitability for fishes and invertebrates under climate change scenarios. *Environmental Management* 55: 836-856.

Kiker, C. F., J. W. Milon, and A. W. Hodges. 2001. Adaptive learning for science based policy: The Everglades restoration. *Ecological Economics* 37: 403-416.

Kline, M., P. E. Frezza, M. R. Robinson, and J. J. Lorenz. 2017. Monitoring Hydrology, Aquatic Vegetation and Fauna in the Southern Everglades: 2015-16 Annual Report. Audubon Florida. Submitted to the SFWMD.

Koch, M. S., S. Schopmeyer, C. Kyhn-Hansen, and C. J. Madden. 2007. Synergistic effects of high temperature and sulfide on tropical seagrass. *Journal of Experimental Marine Biology and Ecology* 341(1): 91-101.

Koch, M., C. Coronado, M. Miller, D. Rudnick, E. Stabenau, R. Hally, and F. Sklar. 2015. Climate change projected effects on coastal foundation communities of the Greater Everglades using a 2060 scenario: Need for a new management paradigm. *Environmental Management* 55: 857-875.

Koebel, J. W., S. G. Bousquin, D. H. Anderson, Z. Welch, M. D. Cheek, H. Chen, B. C. Anderson, R. Baird, T. Beck, A. Brunell, T. Coughlin, and C. Mallison. Chapter 9: Kissimmee River Restoration and Basin Initiatives. 2017 South Florida Environmental Report—Volume I. South Florida Water Management District, West Palm Beach, FL. http://apps.sfwmd.gov/sfwmd/SFER/2017_sfer_final/v1/chapters/v1_ch9.pdf.

Landres, P. B., P. Morgan, and F. J. Swanson. 1999. Overview of the use of natural variability concepts in managing ecological systems. *Ecological Applications* 9(4): 1179-1188.

Lawson, L. L., J. E. Hill, S. Hardin, L. Vilizzi, and G. H. Copp. 2015. Evaluation of the Fish Invasiveness Screening Kit (FISK v2) for peninsular Florida. *Management of Biological Invasions* 6(4): 413-422.

Leeds, J. 2014. Restoration strategies–design and construction status of water quality improvement projects. Chapter 5A in 2014 South Florida Environmental Report, Volume I: The South Florida Environment. West Palm Beach: South Florida Water Management District.

Lempert, R. J., D. G. Groves, S. W. Popper, and S. C. Bankes. 2006. A general, analytic method for generating robust strategies and narrative scenarios. *Management Science* 52(4): 514-528.

Lentz, K. A., and W. A. Dunson. 1998. Water level affects growth of endangered northeastern bulrush, *Scirpus ancistrochaetus* Schuyler. *Aquatic Botany* 60: 213-219.

Light, S., and J. Dineen. 1994. Water control in the Everglades: A historical perspective. In Everglades: The Ecosystem and Its Restoration, S. Davis and J. Ogden, Eds. Delray Beach, FL: St. Lucie Press.

Lindenmayer, D. B., and G. E. Likens. 2009. Adaptive monitoring: A new paradigm for long-term research and monitoring. *Trends in Ecology & Evolution* 24(9): 482-486.

Lord, L. A. 1993. Guide to Florida Environmental Issues and Information. Winter Park: Florida Conservation Foundation.

Marshall, C., Jr., R. Pielke, Sr., L. Steyaert, and D. Willard. 2004. The impact of anthropogenic land cover change on the Florida peninsula sea breezes and warm season sensible weather. *Monthly Weather Review* 132: 28-52.

Mazzei, V., E. E. Gaiser, J. Kominoski, B. J. Wilson, S. Servais, L. Bauman, S. E. Davis, S. P. Kelly, F. H. Sklar, D. T. Rudnick, J. Stachelek, and T. Troxler. 2018. Functional and Compositional Responses of Periphyton Mats to Simulated Saltwater Intrusion in the Southern Everglades. Estuaries and Coasts. doi: 10.1007/s12237-018-0415-6.

McPherson, B. F., and R. Halley. 1996. The South Florida Environment: A Region Under Stress. USGS Circular 1134. Washington, DC: U.S. Government Printing Office.

McVoy, C. W., W. P. Said, J. Obeysekera, J. A. VanArman, and T. W. Dreschel. 2011. Landscapes and Hydrology of the Predrainage Everglades. Gainesville, FL: University of Florida Press.

Meeder, J. F., R. W. Parkinson, P. L. Ruiz, and M. S. Ross. 2017. Saltwater encroachment and prediction of future ecosystem response to the Anthropocene Marine Transgression, Southeast Saline Everglades, Florida. *Hydrobiologia* 803(1): 29-48.

Misra, V., and S. DiNapoli. 2013. Understanding wet season variations over Florida. *Climate Dynamics* 40(5-6): 1361-1372. doi:10.1007/s00382-012-1382-4.

Misra, V., S. DiNapoli, and S. Bastola. 2012a. Dynamic downscaling of the 20th century reanalysis over the southeastern United States. *Regional Environmental Change* 13: S15-23. doi: 10.1007/s10113-012-0372-8.

Misra, V., J. P. Michael, R. Boyles, E. P. Chassignet, M. Griffin, and J. J. O'Brien. 2012b. Reconciling the spatial distribution of the surface temperature trends in the Southeastern United States. *Journal Climate* 25(10): 3610-3618. doi:http://dx.doi.org/10.1175/JCLI-D-11-001701.1.

NASEM (National Academies of Science, Engineering, and Medicine). 2016. Progress Toward Restoring the Everglades: The Sixth Biennial Review—2016. Washington, DC: The National Academies Press.

NASEM. 2017. Effective Monitoring to Evaluate Ecological Restoration in the Gulf of Mexico. Washington, DC: The National Academies Press.
Neubauer, S. C., R. B. Franklin, and D. J. Berrier. 2013. Saltwater intrusion into tidal freshwater marshes alters the biogeochemical processing of organic carbon. *Biogeosciences* 1012: 8171-8183. doi: 10.5194/bg-10-8171-2013.
Nilsson, C., A. L. Aradottir, D. Hagen, G. Halldórsson, K. Høegh, R. J. Mitchell, K. Raulund-Rasmussen, K. Svavarsdóttir, A. Tolvanen, and S. D. Wilson. 2016. Evaluating the process of ecological restoration. *Ecology and Society* 21(1).
Nishihiro, J., M. A. Nishihiro, and I. Washitani. 2006. Assessing the potential for recovery of lakeshore vegetation: Species richness of sediment propagule banks. *Ecological Research* 21(3): 436-445.
NOAA (National Oceanic and Atmospheric Administration). 2014. Sea Level Rise and Nuisance Flood Frequency Changes around the United States. Technical Report NOS CO-OPS 073. Eds. W. V. Sweet, J. Park, J. Marra, C. Zervas, and S. Gill. http://tidesandcurrents.noaa.gov/publications/NOAA_Technical_Report_NOS_COOPS_073.pdf.
NOAA. 2017. Global and Regional Sea Level Rise Scenarios for the United States. Silver Spring, MD: Center for Operational Oceanographic Products and Services. https://tidesandcurrents.noaa.gov/publications/techrpt83_Global_and_Regional_SLR_Scenarios_for_the_US_final.pdf.
NPS (National Park Service). 2012. Guidance for Designing an Integrated Monitoring Program Natural Resource Report NPS/NRSS/NRR—2012/545.
NPS. 2016. Modified Water Deliveries: Improving Hydrologic Conditions in Northeast Shark River Slough. https://www.nps.gov/ever/learn/nature/modwater.htm.
NRC (National Research Council). 1996. Upstream. Washington, DC: National Academy Press.
NRC. 1999. New Directions for Water Resources Planning for the U.S. Army Corps of Engineers. Washington, DC: National Academy Press.
NRC. 2000. Ecological Indicators for the Nation. Washington, DC: National Academies Press.
NRC. 2001. Aquifer Storage and Recovery in the Comprehensive Everglades Restoration Plan: A Critique of the Pilot Projects and Related Plans for ASR in the Lake Okeechobee and Western Hillsboro Areas. Washington, DC: National Academy Press.
NRC. 2002a. Florida Bay Research Programs and Their Relation to the Comprehensive Everglades Restoration Plan. Washington, DC: The National Academies Press.
NRC. 2002b. Regional Issues in Aquifer Storage and Recovery for Everglades Restoration. Washington, DC: The National Academies Press.
NRC. 2003a. Adaptive Monitoring and Assessment for the Comprehensive Everglades Restoration Plan. Washington, DC: The National Academies Press.
NRC. 2003b. Does Water Flow Influence Everglades Landscape Patterns? Washington, DC: The National Academies Press.
NRC. 2003c. Science and the Greater Everglades Ecosystem Restoration: An Assessment of the Critical Ecosystem Studies Initiative. Washington, DC: The National Academies Press.
NRC. 2004. River Basins and Coastal Systems Planning Within the U.S. Army Corps of Engineers. Washington, DC: The National Academies Press.
NRC. 2005. Re-Engineering Water Storage in the Everglades: Risks and Opportunities. Washington, DC: The National Academies Press.
NRC. 2007. Progress Toward Restoring the Everglades: The First Biennial Review—2006. Washington, DC: The National Academies Press.
NRC. 2008. Progress Toward Restoring the Everglades: The Second Biennial Review—2008. Washington, DC: The National Academies Press.
NRC. 2010. Progress Toward Restoring the Everglades: The Third Biennial Review—2010. Washington, DC: The National Academies Press.
NRC. 2012. Progress Toward Restoring the Everglades: The Fourth Biennial Review—2012. Washington, DC: The National Academies Press.
NRC. 2014. Progress Toward Restoring the Everglades: The Fifth Biennial Review—2014. Washington, DC: The National Academies Press.

NRC. 2015. Review of the Everglades Aquifer Storage and Recovery Regional Study. Washington, DC: The National Academies Press.

Nungesser, M., C. Saunders, C. Coronado-Molina, J. Obeysekera, J. Johnson, C. McVoy, and B. Benscoter. 2015. Potential effects of climate change on Florida's Everglades. *Environmental Management* 55(4): 824–835.

Nuttle, W. K., J. W. Fourqurean, B. J. Cosby, J. C. Zieman, and M. B. Robblee. 2000. Influence of net freshwater supply on salinity in Florida Bay. *Water Resources Research* 36(7): 1805-1822.

Obeysekera, J. 2013. Validating climate models for computing evapotranspiration in hydrologic studies: How relevant are climate model simulations over Florida? *Regional Environmental Change* 13(Supplement 1): 81-90.

Obeysekera, J., J. Barnes, and M. Nungesser. 2015. Climate sensitivity runs and regional hydrologic modeling for predicting the response of the greater florida everglades ecosystem to climate change. *Environmental Management* 55: 749-762.

Ogden, J. C., S. M. Davis, K. J. Jacobs, T. Barnes, H. E. Fling, H.E., 2005. The use of conceptual ecological models to guide ecosystem restoration in South Florida. *Wetlands* 25(4): 795-809.

Olsen, J. H., and D. M. Robertson. 2003. Monitoring designs. Seeking a common framework for water quality monitoring. *Water Resources IMPACT* 5(5): 14-16.

Orem, W. H., C. Gilmour, D. Axelrad, D. P. Krabbenhoft, D. Scheidt, P. I. Kalla, P. McCormick, M. Gabriel, and G. Aiken. 2011. Sulfur in the South Florida ecosystem: Distribution, sources, biogeochemistry, impacts, and management for restoration. *Critical Reviews in Environmental Science and Technology* 41(Supplement 1): 249-288.

Park, J., E. Stabenau, J. Redwine, and K. Kotun. 2017. South Florida's encroachment of the sea and environmental transformation over the 21st century. *Journal of Marine Science and Engineering* 5(3): 31.

Pastorok, R. A., A. MacDonald, J. R. Sampson, P. Wilber, D. J. Yozzo, and J. P. Titre. 1997. An ecological decision framework for environmental restoration projects. *Ecological Engineering* 9(1-2): 89-107.

Payne, G. G., and S. K. Xue. 2012. Chapter 3A: Water Quality in the Everglades Protection Area. 2012 South Florida Environmental Report—Volume 1. South Florida Water Management District, West Palm Beach, FL. http://apps.sfwmd.gov/sfwmd/SFER/2012_SFER/v1/chapters/v1_ch3a.pdf.

Perry, W. 2004. Elements of South Florida's Comprehensive Everglades Restoration Plan. *Ecotoxicology* 13: 185-193.

Phlips, E. J., F. J. Aldridge, P. Hansen, P. V. Zimba, J. Ihnat, M. Conroy, and P. Ritter. 1993a. Spatial and temporal variability of trophic state parameters in a shallow subtropical lake (Lake Okeechobee, Florida, USA). *Archiv für Hydrobiologie* 128: 437-458.

Phlips, E. J., P. V. Zimba, M. S. Hopson, and T. L. Crisman. 1993b. Dynamics of the plankton community in submerged plant dominated regions of Lake Okeechobee, Florida, USA. *Internationale Vereinigung für theoretische und angewandte Limnologie: Verhandlungen* 25(1): 423-426.

Phlips, E. J., S. Badylak, J. Hart, D. Haunert, J. Lockwood, K. O'Donnell, D. Sun, P. Viveros, and M. Yilmaz. 2012. Climatic influences on autochthonous and allochthonous phytoplankton blooms in a subtropical estuary, St. Lucie Estuary, Florida, USA. *Estuaries Coasts* 35:335-352.

Piccone, T. T. 2010. Chapter 8: Implementation of the Long-term Plan for Achieving Water Quality Goals in the Everglades Protection Area. 2010 South Florida Environmental Report—Volume 1. South Florida Water Management District, West Palm Beach, FL. https://my.sfwmd.gov/portal/page/portal/pg_grp_sfwmd_sfer/portlet_sfer/tab2236037/2010%20report/v1/chapters/v1_ch8.pdf.

Piccone, T. T. 2011. Chapter 8: Implementation of the Long-term Plan for Achieving Water Quality Goals in the Everglades Protection Area. 2011 South Florida Environmental Report—Volume 1. South Florida Water Management District, West Palm Beach, FL. http://my.sfwmd.gov/portal/page/portal/pg_grp_sfwmd_sfer/portlet_prevreport/2011_sfer/v1/chapters/v1_ch8.pdf.

Prato, T. 2005. Bayesian adaptive management of ecosystems. *Ecological Modelling* 183(2-3): 147-156.

Qui, C. 2016. Volume III. Appendix 2-4: Annual Permit Report for the C-111 Spreader Canal Phase 1 (Western) Project. 2016 South Florida Environmental Report. Available at http://apps.sfwmd.gov/sfwmd/SFER/2016_sfer_final/v3/appendices/v3_app2-4.pdf (accessed November 2, 2016).

Qui, C., J. Godin, B. Gu, and J. Shaffer. 2018. Appendix 2-4: Annual Permit Report for the C-111 Spreader Canal Phase 1 (Western) Project. 2016 South Florida Environmental Report—Volume III. http://apps.sfwmd.gov/sfwmd/SFER/2018_sfer_final/v3/appendices/v3_app2-4.pdf.

Rayer, S., and Y. Wang. 2018. Projections of Florida Population by County, 2020–2045, with Estimates for 2017. Bureau of Economic and Business Research. Vol. 51, Bulletin 180. https://www.bebr.ufl.edu/sites/default/files/Research%20Reports/projections_2018.pdf.

RECOVER. 2005. The Recover Team's Recommendations for Interim Goals and Interim Targets for the Comprehensive Everglades Restoration Plan. West Palm Beach, FL: RECOVER.

RECOVER. 2006. Monitoring and Assessment Plan (MAP) Part 2: 2006 Assessment Strategy for the MAP. http://141.232.10.32/pm/recover/recover_map_part2.aspx.

RECOVER. 2009. 2009 RECOVER Monitoring and Assessment Plan. Jacksonville, FL: U.S. Army Corps of Engineers and West Palm Beach: South Florida Water Management District. http://141.232.10.32/pm/recover/recover_map_2009.aspx.

RECOVER. 2010. Final RECOVER 2009 System Status Report. September 2010. Jacksonville, FL: U.S. Army Corps of Engineers and West Palm Beach: South Florida Water Management District. http://141.232.10.32/pm/ssr_2009/ssr_pdfs/2009_ssr_full_web.pdf.

RECOVER. 2014. 2014 System Status Report. August 2014. Jacksonville, FL: U.S. Army Corps of Engineers and West Palm Beach: South Florida Water Management District, http://141.232.10.32/pm/ssr_2014/cerp_ssr_2014.aspx; Chapter 7, Southern Coastal Systems (includes PSRP, C-111 SC, and BBCW), http://141.232.10.32/pm/ssr_2014/Docs/mod_scs_2014.pdf.

RECOVER. 2015. Program-Level Adaptive Management Plan: Comprehensive Everglades Restoration Plan. September 8, 2015. http://www.saj.usace.army.mil/Portals/44/docs/Environmental/RECOVER/20151019_CERPPROGRAMAMPLAN_DCT_APPROVED.pdf.

RECOVER. 2016. Restoration Coordination and Verification Five Year Plan: A Plan to Support the Changing Needs of the Comprehensive Everglades Restoration Plan, Fiscal Years 2017-2021.

Reynolds, J. H., M. G. Knutson, K. B. Newman, E. D. Silverman, and W. L. Thompson. 2016. A road map for designing and implementing a biological monitoring program. *Environmental Monitoring and Assessment* 188: 399.

Reichert, B. E., W. L. Kendall, R. J. Fletcher, and W. E. Kitchens. 2016. Spatio-temporal variation in age structure and abundance of the endangered snail kite: Pooling across regions masks a declining and aging population. PLoS ONE. http://dx.doi.org/10.1371/journal.pone.0162690.

Richardson, J. R., and T. T. Harris. 1995. Vegetation mapping and change detection in the Lake Okeechobee marsh ecosystem. *Archiv für Hydrobiologie, Advances in Limnology* 45: 17-39.

Rivera-Monroy, V. H., K. de Mutsert, R. R. Twilley, E. Castañeda-Moya, M. M. Romigh, and S. E. Davis. 2007. Patterns of nutrient exchange in a riverine mangrove forest in the Shark River Estuary, Florida, USA. *Hydrobiologica* 17(2): 169-178.

Robinson, M. R., M. Kline, P. E. Frezza, and J. J. Lorenz. 2016. Monitoring Hydrology, Aquatic Vegetation and Fauna in the Southern Everglades: 2014-15 Annual Report. Audubon Florida. Submitted to the SFWMD.

Ross, M. S., J. F. Meeder, J. P. Sah, P. L. Ruiz, and G. J. Telesnicki. 2000. The Southeast Saline Everglades revisited: 50 years of coastal vegetation change. *Journal of Vegetation Science* 11(1): 101-112.

Salas, J. D., J. Obeysekera, and R. M. Vogel. 2018. Techniques for assessing water infrastructure for nonstationary extreme events: A review. *Hydrological Sciences Journal* 63(3). https://doi.org/10.1080/02626667.2018.1426858.

Salathé, E. P., Jr., P. W. Mote, and M. W. Wiley. 2007. Review of scenario selection and downscaling methods for the assessment of climate change impacts on hydrology in the United States Pacific Northwest. *International Journal of Climatology* 27: 1611-1621.

Sas, H. 1989. Lake Restoration by Reduction of Nutrient Loadings: Expectations, Experiences, Extrapolations. Germany: Academia Verlag Richarz.

Scheffer, M. 1989. Alternative stable states in eutrophic shallow freshwater systems: A minimal model. *Hydrobiological Bulletin* 23: 73-85.

Scheffer, M., S. Carpenter, J. Foley, C. Folke, and B. Walker. 2001. Catastrophic shifts in ecosystems. *Nature* 413: 591-596.

SCT (Science Coordination Team). 2003. The Role of Flow in the Everglades Ridge and Slough Landscape. http://www.sfrestore.org/sct/docs/.

Selman, C., V. Misra, L. Stefanova, S. DiNapoli, and T. J. Smith III. 2013. On the twenty-first-century wet season projections over the Southeastern United States. *Regional Environmental Change* 13(Supplement 1): 153-164.

SFERTF (South Florida Ecosystem Restoration Task Force). 2000. Coordinating Success: Strategy for Restoration of the South Florida Ecosystem. July. http://www.sfrestore.org/documents/work_products/coordinating_success_2000.pdf.

SFERTF. 2010. Plan for Coordinating Science Update: July 2008-June 2010.

SFERTF. 2016. South Florida Integrated Financial Plan. https://evergladesrestoration.gov/content/documents/integrated_financial_plan/2016/2016_Integrated_Financial_Plan.pdf.

SFERTF. 2018. 2018 Cross Cut Budget. https://evergladesrestoration.gov/content/documents/cross_cut_budget/2018/2018_cross_cut_budget.pdf.

SFWMD (South Florida Water Management District). 2011. Past and Project Trends in Climate and Sea Level for South Florida. South Florida Water Management District, West Palm Beach, FL. https://www.sfwmd.gov/sites/default/files/documents/ccireport_publicationversion_14jul11.pdf.

SFWMD. 2013. C-111 Spreader Canal Western Project: Just the Facts. January 2013.

SFWMD. 2017. Restoration Strategies and CERP Project Updates: Lower West Coast Water Supply Plan Meeting. March 23, 2017. https://www.sfwmd.gov/sites/default/files/documents/lwc_2017_plan_032317_pres_everglades_restoration.pdf.

SFWMD. 2018a. Central Everglades Planning Project Post Authorization Change Report: Feasibility Study and Draft Environmental Impact. March 2018. South Florida Water Management District, West Palm Beach, FL. https://www.sfwmd.gov/sites/default/files/documents/cepp_pacr_main_report.pdf.

SFWMD. 2018b. Central Everglades Planning Project Post Authorization Change Report: Feasibility Study and Draft Environmental Impact Dated March 2018: Addendum, May 2018. https://www.sfwmd.gov/sites/default/files/documents/Addendum_SFWMD SEC 203 CEPP PACR (05.25.2018).pdf.

SFWMD. 2018c. South Florida Environmental Report. https://www.sfwmd.gov/science-data/scientific-publications-sfer.

Skalski, J. R., D. A. Coats, and A. K. Fukuyama. 2001. Criteria for oil spill recovery: A case study of the intertidal community in Prince William Sound, Alaska, following the Exxon Valdez oil spill. *Environmental Management* 28(1): 9-18.

Sklar, F. 2013. An Update of the Decomp. Physical Model (DPM): The Largest AM Assessment in USACOE History. Joint Working Group and Science Coordination Group Meeting November 19. http://evergladesrestoration.gov/content/wg/minutes/2013meetings/111913/Decompartmentalization_Physical_Model_Update.pdf.

Sklar, F., and T. Dreschel. 2018. Chapter 6: Everglades Research and Evaluation. 2018 South Florida Environmental Report—Volume 1. http://apps.sfwmd.gov/sfwmd/SFER/2018_sfer_final/v1/chapters/v1_ch6.pdf.

Sloey, T. M., R. J. Howard, and M. W. Hester. 2016. Response of *Schoenoplectus acutus* and *Schoenoplectus californicus* at different life history stages to hydrologic regime. *Wetlands* 36: 37-46.

Smith, D. H., and M. Smart. 2005. Influence of water depth on persistence of giant bulrush communities in Lake Okeechobee, Florida. Report to the South Florida Water Management District.

Smith, J. P., and M. W. Collopy. 1995. Colony turnover, nest success and productivity, and causes of nest failure among wading birds (Ciconiiformes) at Lake Okeechobee, Florida (1989-2002). *Archiv für Hydrobiologie, Advances in Limnology* 45: 287-316.

Smith, J. P., J. R. Richardson, and M. W. Collopy. 1995. Foraging habitat selection among wading birds (Ciconiiformes) at Lake Okeechobee, Florida, in relation to hydrology and vegetative cover. *Archiv für Hydrobiologie, Advances in Limnology* 45: 247-285.

Smoak, J. M., J. L. Breithaupt, T. J. Smith III, and C. J. Sanders. 2013. Sediment accretion and organic carbon burial relative to sea-level rise and storm events in two mangrove forests in Everglades National Park. *Catena* 104: 58-66. doi: 10.1016/j.catena.2012.10.009.

Society for Ecological Restoration International Science & Policy Working Group. 2004. The SER International Primer on Ecological Restoration. Tucson, AZ: Society for Ecological Restoration International.

Sondergaard, M., P. Kristensen, and E. Jeppesen. 1992. Phosphorous release from resuspended sediment in the shallow and wind-exposed Lake Arreso, Denmark. *Hydrobiologia* 228: 91-99.

Southeast Florida Regional Climate Change Compact Sea Level Rise Work Group (Compact). October 2015. Unified Sea Level Rise Projection for Southeast Florida. A document prepared for the Southeast Florida Regional Climate Change Compact Steering Committee. 35 pp.

Spalding, M. D., A. L. McIvor, M. W. Beck, E. W. Koch, I. Möller, D. J. Reed, P. Rubinoff, T. Spencer, T. J. Tolhurst, T. V. Wamsley, B. K. van Wesenbeeck, E. Wolanski, and C. D. Woodroffe. 2014. Coastal ecosystems: A critical element of risk reduction. *Conservation Letters* (7):293–301.

Squires, L., and A. G. Van Der Valk. 1992. Water-depth tolerances of the dominant emergent macrophytes of the Delta Marsh, Manitoba. *Canadian Journal of Botany* 70: 1860-1867.

SSG (Science Sub-Group). 1993. Federal Objectives for the South Florida Restoration by the Science Sub-Group of the South Florida Management and Coordination Working Group. Miami, FL.

Steinman, A. D., K. E. Havens, A. J. Rodusky, B. Sharfstein, R. T. James, and M. C. Harwell. 2002. The influence of environmental variables and a managed water recession on the growth of charophytes in a large subtropical lake. *Aquatic Botany* 72: 297-313.

Tebaldi, C., R. L. Smith, D. Nychka, and L. O. Mearns. 2005. Quantifying uncertainty in projections of regional climate change: A Bayesian approach to the analysis of multimodel ensembles. *Journal of Climate* 18(10):1524-1540.

Thom, R., and K. Wellman. 1996. Planning Aquatic Ecosystem Restoration Monitoring Programs. IWR Report 96-R-23. U.S. Army Corps of Engineers Institute for Water Resources. Alexandria, VA.

Thomas, M. L., N. Gunawardene, K. Horton, A. Williams, S. O'Connor, S. McKirdy, and J. van der Merwe. 2017. Many eyes on the ground: Citizen science is an effective early detection tool for biosecurity. *Biological Invasions* 19(9): 2751-2765.

Titus, J. G., and C. Richman. 2001. Maps of lands vulnerable to sea level rise: Modeled elevations along the US Atlantic and Gulf coasts. *Climate Research* 183: 205-228.

Troxler, T. G., D. L. Childers, and C. J. Madden. 2013. Drivers of decadal-scale change in Southern Everglades wetland macrophyte communities of the coastal ecotone. *Wetlands* 34(1): 81-90.

USACE (U.S. Army Corps of Engineers). 1992. General Design Memorandum and Environmental Impact Statement: Modified Water Deliveries to Everglades National Park. Atlanta, GA: USACE.

USACE. 2007a. Memorandum for Director of Civil Works on Comprehensive Everglades Restoration Plan, Water Quality Improvements. Washington, DC: USACE.

USACE. 2007b. Final Supplemental Environmental Impact Statement: Lake Okeechobee Regulation Schedule. Jacksonville, FL: USACE.

USACE. 2008. Water Control Plan for Lake Okeechobee and Everglades Agricultural Area. USACE, Jacksonville District.

USACE. 2009. CECW-PB. Implementation Guidance for Section 2039 of the Water Resources Development Act of 2007 (WRDA 2007)—Monitoring Ecosystem Restoration. Memorandum for Commanders. https://cw-environment.erdc.dren.mil/restore/pdfs/Section%202039%20of%20 WRDA%202007%20Monitoring%20Ecosystem%20Restoration.pdf.

USACE. 2013. Indian River Lagoon-South. Facts and Information. Jacksonville, FL.

USACE. 2015a. South Dade Investigation Workshop, Meeting Presentation. Homestead, FL. October 15, 2015. https://www.sfwmd.gov/sites/default/files/documents/sdi_2015_10_15_usace_george_pres.pdf.

USACE. 2015b. December 2015. C-111 South Dade Project. Facts and Information.

USACE. 2016a. South Florida Ecosystem Restoration (SFER) Program Overview. http://www.saj.usace.army.mil/Portals/44/docs/Environmental/Everglades%20Restoration%20Overview%20Placemat_May2016_web.pdf.

USACE. 2016b. (January 27). G-3273 and S-356 Pump Station Field Test. Presentations from meeting of Project Delivery Team. West Palm Beach.

USACE 2016c. Ecological Monitoring Plan for the C-111 Spreader Canal Western Project, revised January 25, 2016.

USACE. 2016d. Herbert Hoover Dike Rehabilitation Project. PowerPoint Presentation, U.S. Army Corps of Engineers, Jacksonville District.

USACE. 2017. Modified Water Deliveries to Everglades National Park: G-3273 & S-356 Pump Station Field Test. Facts and Information. https://usace.contentdm.oclc.org/utils/getfile/collection/p16021coll11/id/2243.

USACE. 2018a. News Release: President's Fiscal 2019 Budget for U.S. Army Corps of Engineers Civil Works Program Released. Release no. 18-015. https://www.usace.army.mil/Media/News-Releases/News-Release-Article-View/Article/1438488/presidents-fiscal-2019-budget-for-us-army-corps-of-engineers-civil-works-progra/.

USACE. 2018b. Integrated Delivery Schedule (IDS) SFER Program Snapshot Through 2030. July 2018 Update. U.S. Army Corps of Engineers—Jacksonville District. https://usace.contentdm.oclc.org/utils/getfile/collection/p16021coll11/id/2641.

USACE. 2018c. Site 1 Impoundment Facts and Information. U.S. Army Corps of Engineers—Jacksonville District. https://usace.contentdm.oclc.org/utils/getfile/collection/p16021coll11/id/2583.

USACE. 2018d. Biscayne Bay: Coastal Wetlands Projects. Fact Sheet. http://cdm16021.contentdm.oclc.org/utils/getfile/collection/p16021coll11/id/2015.

USACE. 2018e. C-111 Spreader Canal Western Project—Fact Sheet. https://usace.contentdm.oclc.org/utils/getfile/collection/p16021coll11/id/2252.

USACE. 2018f. Facts & Information: Modified Water Deliveries to Everglades National Park. https://usace.contentdm.oclc.org/utils/getfile/collection/p16021coll11/id/2191.

USACE. 2018g. Herbert Hoover Dike Rehabilitation: Project Update. Fact Sheet. Summer 2018.

USACE. 2018h. $3.348 Billion in Recovery Funds Go Toward Reducing Flood Risk in Florida, Puerto Rico and U.S. Virgin Islands. Release no. 18-047. July 5, 2018.

USACE and DOI. 2016. Central and Southern Florida Project, Comprehensive Everglades Restoration Plan, Report to Congress, 2015. http://www.saj.usace.army.mil/Portals/44/docs/Environmental/Report%20to%20Congress/FINAL_RTC_2015_01Mar16fin-WithLetters-WithCovers-508Compliant.pdf.

USACE and SFWMD. 1999. Central and Southern Florida Project Comprehensive Review Study, Final Integrated Feasibility Report and Programmatic Environmental Impact Statement. U.S. Army Corps of Engineers and South Florida Water Management District.

USACE and SFWMD. 2004a. Comprehensive Everglades Restoration Plan Picayune Strand Restoration (Formerly Southern Golden Gate Estates Ecosystem Restoration) Final Integrated Project Implementation Report and Environmental Impact Statement.

USACE and SFWMD. 2004b. Central and Southern Florida Project Comprehensive Everglades Restoration Plan: Conceptual Alternatives, Everglades Agricultural Area Storage Reservoirs Phase 1. http://141.232.10.32/pm/projects/project_docs/pdp_08_eaa_store/040404_docs_08_conceptual_alt.pdf.

USACE and SFWMD. 2009. Picayune Strand Restoration Project, Annex I to the Transfer Agreement: Monitoring Plan. August 2009.

USACE and SFWMD. 2010. Caloosahatchee River (C-43) West Basin Storage Reservoir Final Integrated Project Implementation Report and Environmental Impact Statement. http://141.232.10.32/pm/projects/docs_04_c43_pir_final.aspx.

USACE and SFWMD. 2011a. CERP Guidance Memorandum 56: Guidance for Integration of Adaptive Management into Comprehensive Everglades Restoration Plan Program and Project Management. February 8, 2011. http://www.cerpzone.org/documents/cgm/CGM_56_Adaptive_Management.pdf.

USACE and SFWMD. 2011b. Biscayne Bay Coastal Wetlands Phase 1 Final Integrated Project Implementation Report and Environmental Impact Statement. July 2011. Annex E: Project Monitoring Plan. http://141.232.10.32/pm/projects/project_docs/pdp_28_biscayne/010612_fpir/010612_vol_3_annex_e.pdf.

USACE and SFWMD. 2012. Central and Southern Florida Project Comprehensive Everglades Restoration Plan Biscayne Bay Coastal Wetlands Phase 1: Final Integrated Project Implementation Report and Environmental Impact Statement. July 2011—Revised March 2012.

USACE and SFWMD. 2014. Central and Southern Florida Project Comprehensive Everglades Restoration Plan Central Everglades Planning Project: Final Integrated Project Implementation Report and Environmental Impact Statement. http://141.232.10.32/docs/2014/08/01_CEPP%20Final%20PIR-EIS%20Main%20Report.pdf.

USACE and SFWMD. 2015a. Melaleuca Eradication and Other Exotic Plants: Implement Biological Control. Facts and Information.

USACE and SFWMD. 2015b. Comprehensive Everglades Restoration Plan: Aquifer Storage and Recovery Regional Study: Technical Data Report. May 2015. U.S. Army Corps of Engineers, Jacksonville, FL, and South Florida Water Management District, West Palm Beach, FL.

USACE and SFWMD. 2016a (February). Site 1 Impoundment/Fran Reich Preserve: Facts and Information. http://www.saj.usace.army.mil/Portals/44/docs/FactSheets/Site1_FS_February2016_web.pdf.

USACE and SFWMD. 2016b. C-43 West Basin Storage Reservoir Project. Facts and Information.

USACE and SFWMD. 2018a. Loxahatchee River: Watershed Restoration Project – Facts and Information. http://cdm16021.contentdm.oclc.org/utils/getfile/collection/p16021coll11/id/2048.

USACE and SFWMD. 2018b. Presentation to the Lake Okeechobee Watershed Restoration Project (LOWRP) Project Delivery Team. May 2, 2018.

USACE and SFWMD. 2018c. Lake Okeechobee Watershed Restoration Project: Draft Integrated Project Implementation Report and Environmental Impact Statement. https://usace.contentdm.oclc.org/utils/getfile/collection/p16021coll7/id/7413.

USACE, DOI, and the State of Florida. 2007. Intergovernmental Agreement Among the United States Department of the Army, the United States Department of the Interior, and the State of Florida Establishing Interim Restoration Goals for the Comprehensive Everglades Restoration Plan. http://141.232.10.32/pm/pm_docs/prog_regulations/081607_int_goals.pdf.

USDA (U.S. Department of Agriculture). 2003. Plant Guide: California Bulrush. https://plants.sc.egov.usda.gov/plantguide/pdf/cs_scca11.pdf.

USGCRP (U.S. Global Change Research Program). 2017. Climate Science Special Report: Fourth National Climate Assessment, Volume I. D. J. Wuebbles, D. W. Fahey, K. A. Hibbard, D. J. Dokken, B. C. Stewart, and T. K. Maycock, Eds. U.S. Global Change Research Program, Washington, DC, 470 pp., doi: 10.7930/J0J964J6.

USGS (U.S. Geological Survey) and Delta Stewardship Council. 2018. The Science Enterprise Workshop: Supporting and Implementing Collaborative Science Excutive Summary. April 16, 2018. http://deltacouncil.ca.gov/docs/delta-plan-implementation-committee/science-enterprise-workshop-supporting-and-implementing.

Valle-Levinson, A., A. Dutton, and J. B. Martin. 2017. Spatial and temporal variability of sea level rise hot spots over the eastern United States. *Geophysical Research Letters* 44(15).

Volety, A. K., and S. G. Tolley. 2005. Life History and Ecological Aspects of the American (Eastern) Oyster, *Crassostrea virginica*. Technical Report. Southwest Florida Water Management District, 49 pp.

Volety, A. K., M. Savarese, S. G. Tolley, W. S. Arnold, P. Sime, P. Goodman, R. H. Chamberlain, and P.H. Doering. 2009. Eastern oysters (*Crassostrea virginica*) as an indicator for restoration of Everglades Ecosystems. *Ecological Indicators* 9S: 120-136.

Volk, M. I., T. S. Hoctor, B. B. Nettles, R. Hilsenbeck, F. E. Putz, and J. Oetting. 2017. Florida Land Use and Land Cover Change in the Past 100 Years. Florida's Climate: Changes, Variations, & Impacts (Edition 1).

Vymazal, J. 2011. Plants used in constructed wetlands with horizontal subsurface flow: A review. *Hydrobiologia* 647:133-156.

Wachnicka, A. H., and G. L. Wingard. 2015. Biological indicators of changes in water quality and habitats of the coastal and estuarine areas of the Greater Everglades ecosystem. In J.A. Entry, A.D. Gottlieb, K. Jayachandrahan, and A. Ogram (editors), Microbiology of the Everglades Ecosystem. Boca Raton, FL: CRC Press, pp. 218-240.

Wanless, H. R., and M. G. Tagett. 1989. Origin, growth and evolution of carbonate mudbanks in Florida Bay. *Bulletin of Marine Science* 44(1): 454-489.

Wanless, H. R., and B. M. Vlaswinkel. 2005. Coastal Landscape and Channel Evolution Affecting Critical Habitats at Cape Sable, Everglades National Park, Florida. Final report to Everglades National Park National Park Service, U.S. Department of Interior. 197 pp.

Warszawski, L., A. Friend, S. Ostberg, K. Frieler, W. Lucht, S. Schaphoff, D. Beerling, P. Cadule, P. Ciais, D. B. Clark, and R. Kuhana. 2013. A multi-model analysis of risk of ecosystem shifts under climate change. *Environmental Research Letters* 8(4): 044018.

Wilson, B. J., S. Servais, S. P. Charles, S. E. Davis, E. E. Gaiser, J. Kominoski, J. H. Richards, and T. Troxler. 2018. Declines in plant productivity drive carbon loss from brackish coastal wetland mesocosms exposed to saltwater intrusion. *Estuaries and Coasts* doi: 10.1007/s12237-018-0438-z.

Wingard, G. L., and J. J. Lorenz. 2014. Integrated conceptual ecological model and habitat indices for the southwest Florida coastal wetlands. *Ecological Indicators* 44: 92-107. https://pubs.er.usgs.gov/publication/70156769.

Wingard, G. L., C. E. Bernhardt, and A. H. Wachnicka. 2017. The role of paleoecology in restoration and resource management—The past as a guide to future decision-making: Review and example from the greater Everglades ecosystem, U.S.A. *Frontiers in Ecology and Evolution* 5(11). doi: 10.3389/fevo.2017.00011.

Wood, A. W., L. R. Leung, V. Sridhar, and D. P. Lettenmaier. 2004. Hydrologic implications of dynamical and statistical approaches to downscaling climate model output, *Climatic Change* 62: 189-216.

Worley, K. B., J. R. Schmid, M. J. Schuman, V. G. Booher, L. A. Johnson, D. Addison, and I. A. Bartoszek. 2017. Attachment E: First Year Post-Restoration Aquatic Fauna Monitoring in the Picayune Strand Restoration Project Area (2016-2017). Conservancy of Southwest Florida. Appendix 2-1: Annual Permit Report for the Picayune Strand Restoration Project. 2018 South Florida Environmental Report Volume III. SFWMD. http://apps.sfwmd.gov/sfwmd/SFER/2018_sfer_final/v3/appendices/v3_app2-1.pdf.

Zhang, J., and Z. Welch. 2018. Chapter 8B: Lake Okeechobee Watershed Research and Water Quality Monitoring Results and Activities. 2018 South Florida Environmental Report – Volume 1. South Florida Water Management District, West Palm Beach, FL.

Appendix A

The National Academies of Sciences, Engineering, and Medicine Everglades Reports

Progress Toward Restoring the Everglades: The Sixth Biennial Review, 2016 (2016)

The 2016 biennial report finds that 16 years into the Comprehensive Everglades Restoration Project (CERP) completed components of the project are beginning to show ecosystem benefits, but the committee had several concerns regarding progress. There has been insufficient attention to refining long-term systemwide goals and objectives and the need to adapt the CERP to radically changing system and planning constraints. It now is known that the natural system was historically much wetter than previously assumed, bringing into question some of the hydrological goals embedded in the restoration plan. Sea-level rise will reduce the footprint of the system, temperature and evaporative water losses will increase, rainfall may become more variable, and more storage will likely be needed to accommodate future increases or decreases in the quantity and intensity of runoff.

Review of the Everglades Aquifer Storage and Recovery Regional Study (2015)

The Florida Everglades is a large and diverse aquatic ecosystem that has been greatly altered over the past century by an extensive water control infrastructure designed to increase agricultural and urban economic productivity. The CERP, launched in 2000, is a joint effort led by the state and federal government to reverse the decline of the ecosystem. Increasing water storage is a critical component of the restoration, and the CERP included projects that would drill more than 330 aquifer storage and recovery (ASR) wells to store up to 1.65 billion gallons per day in porous and permeable units in the aquifer system during wet periods for recovery during seasonal or longer-term dry periods.

To address uncertainties regarding regional effects of large-scale ASR implementation in the Everglades, the U.S. Army Corps of Engineers (USACE) and the

South Florida Water Management District conducted an 11-year ASR Regional Study, with focus on the hydrogeology of the Floridan aquifer system, water quality changes during aquifer storage, possible ecological risks posed by recovered water, and the regional capacity for ASR implementation. At the request of the USACE, this report reviews the ASR Regional Study Technical Data Report and assesses progress in reducing uncertainties related to full-scale CERP ASR implementation. This report considers the validity of the data collection and interpretation methods; integration of studies; evaluation of scaling from pilot- to regional-scale application of ASR; and the adequacy and reliability of the study as a basis for future applications of ASR.

Progress Toward Restoring the Everglades: The Fifth Biennial Review, 2014 (2014)

This report is the fifth biennial evaluation of progress being made in the CERP. Despite exceptional project planning accomplishments, over the past 2 years progress toward restoring the Everglades has been slowed by frustrating financial and procedural constraints. The Central Everglades Planning Project is an impressive strategy to accelerate Everglades restoration and avert further degradation by increasing water flow to the ecosystem. However, timely authorization, funding, and creative policy and implementation strategies will be essential to realize important near-term restoration benefits. At the same time, climate change and the invasion of nonnative plant and animal species further challenge the Everglades ecosystem. The impacts of changing climate—especially sea-level rise—add urgency to restoration efforts to make the Everglades more resilient to changing conditions.

Progress Toward Restoring the Everglades: The Fourth Biennial Review, 2012 (2012)

The 2012 biennial report finds that, 12 years into the Comprehensive Everglades Restoration Project, little progress has been made in restoring the core of the remaining Everglades ecosystem; instead, most project construction so far has occurred along its periphery. To reverse ongoing ecosystem declines, it will be necessary to expedite restoration projects that target the central Everglades, and to improve both the quality and quantity of the water in the ecosystem. The new Central Everglades Planning Project offers an innovative approach to this challenge, although additional analyses are needed at the interface of water quality and water quantity to maximize restoration benefits within existing legal constraints.

Progress Toward Restoring the Everglades: The Third Biennial Review, 2010 (2010)

The 2010 biennial report finds that while natural system restoration progress from CERP remains slow, in the past 2 years, there have been noteworthy improvements in the pace of implementation and in the relationship between the federal and state partners. Continued public support and political commitment to long-term funding will be needed for the restoration plan to be completed. The science program continues to address important issues, but more transparent mechanisms for integrating science into decision making are needed. Despite such progress, several important challenges related to water quality and water quantity have become increasingly clear, highlighting the difficulty of achieving restoration goals simultaneously for all ecosystem components. Achieving these goals will be enormously costly and will take decades at least. Rigorous scientific analyses of potential conflicts among the hydrologic requirements of Everglades landscape features and species, and the tradeoffs between water quality and quantity, considering timescales of reversibility, are needed to inform future prioritization and funding decisions. Understanding and communicating these tradeoffs to stakeholders are critical.

Progress Toward Restoring the Everglades: The Second Biennial Review, 2008 (2008)

The report concludes that budgeting, planning, and procedural matters are hindering a federal and state effort to restore the Florida Everglades ecosystem, which is making only scant progress toward achieving its goals. Good science has been developed to support restoration efforts, but future progress is likely to be limited by the availability of funding and current authorization mechanisms. Despite the accomplishments that lay the foundation for CERP construction, no CERP projects have been completed to date. To begin reversing decades of decline, managers should address complex planning issues and move forward with projects that have the most potential to restore the natural ecosystem.

Progress Toward Restoring the Everglades: The First Biennial Review, 2006 (2007)

This report is the first in a congressionally mandated series of biennial evaluations of the progress being made by the CERP. The report finds that progress has been made in developing the scientific basis and management structures needed to support a massive effort to restore the Florida Everglades ecosystem. However, some important projects have been delayed due to several factors including budgetary restrictions and a project planning process that can be stalled by unresolved scientific uncertainties. The report outlines an alternative approach that can help the initiative move forward even as it resolves remaining scientific uncertainties. The report calls for a boost in the rate of federal spending if the restoration of Everglades National Park and other projects are to be completed on schedule.

Re-engineering Water Storage in the Everglades: Risks and Opportunities (2005)

Human settlements and flood control structures have significantly reduced the Everglades, which once encompassed more than 3 million acres of slow-moving water enriched by a diverse biota. The CERP was formulated in 1999 with the goal of restoring the original hydrologic conditions of the remaining Everglades. A major feature of this plan is providing enough storage capacity to meet human and ecological needs. This report reviews and evaluates not only storage options included in the plan, but also other options not considered in the plan. Along with providing hydrologic and ecological analyses of the size, location, and functioning of water storage components, the report also discusses and makes recommendations on related critical factors, such as timing of land acquisition, intermediate states of restoration, and tradeoffs among competing goals and ecosystem objectives.

The CERP imposes some constraints on sequencing of its components. The report concludes that two criteria are most important in deciding how to sequence components of such a restoration project: (1) protecting against additional habitat loss by acquiring or protecting critical lands in and around the Everglades and (2) providing ecological benefits as early as possible.

There is a considerable range in the degree to which various proposed storage components involve complex design and construction measures, rely on active controls and frequent equipment maintenance, and require fossil fuels or other energy sources for operation. The report recommends that, to the extent possible, the CERP should develop storage components that have fewer of those requirements and are thus less vulnerable to failure and more likely to be sustainable in the long term.

Further, as new information becomes available and as the effectiveness and feasibility of various restoration components become clearer, some of the earlier adaptation and compromises might need to be revisited. The report recommends that methods be developed to allow for assessment of tradeoffs over broad spatial and long temporal scales, especially for the entire ecosystem, and gives an example of what an overall performance indicator for the Everglades system might look like.

Adaptive Monitoring and Assessment for the Comprehensive Everglades Restoration Plan (2003)

A key premise of the CERP is that restoring the historical hydrologic regime in the remaining wetlands will reverse declines in many native species and biological communities. Given the uncertainties that will attend future responses of Everglades ecosystems to restored water regimes, a research, monitoring, and adaptive management program is planned. This report assessed the extent to which the restoration effort's "monitoring and assessment plan" included the following elements crucial to any adaptive management scheme: (1) clear restoration goals and targets, (2) a sound baseline description and conceptualization of the system, (3) an effective process for learning from management actions, and (4) feedback mechanisms for improving management based on the learning process.

The report concludes that monitoring needs must be prioritized, because many goals and targets that have been agreed to may not be achievable or internally consistent. Priorities could be established based on the degree of flexibility or reversibility of a component and its potential impact on future management decisions. Such a prioritization should be used for scheduling and sequencing of projects, for example. Monitoring that meets multiple objectives (e.g., adaptive management, regulatory compliance, and a "report card") should be given priority.

Ecosystem-level, systemwide indicators should be developed, such as land-cover and land-use measures, an index of biotic integrity, and diversity measures. Regionwide monitoring of human and environmental drivers of the ecosystem, especially population growth, land-use change, water demand, and sea-level rise are recommended. Monitoring, modeling, and research should be well integrated, especially with respect to defining the restoration reference state and using "active" adaptive management.

Does Water Flow Influence Everglades Landscape Patterns? (2003)

A commonly stated goal of the CERP is to "get the water right." This has largely meant restoring the timing and duration of water levels and the water quality in the Everglades. Water flow (speed, discharge, direction) has been considered mainly in the coastal and estuarine system, but not elsewhere. Should the restoration plan be setting targets for flows in other parts of the Everglades as well?

There are legitimate reasons why flow velocities and discharges have thus far not received greater emphasis in the plan. These include a relative lack of field information and poor resolution of numerical models for flows. There are, however, compelling reasons to believe that flow has important influences in the central Everglades ecosystem. The most important reason is the existence of major, ecologically important landforms—parallel ridges, sloughs, and "tree islands"—are aligned with present and inferred past flow directions. There are difficulties in interpreting this evidence, however, as it is essentially circumstantial and not quantitative.

Alternative mechanisms by which flow may influence this landscape can to some extent be evaluated from short-term research on underlying bedrock topography, detailed surface topographic mapping, and accumulation rates of suspended organic matter. Nonetheless, more extensive and long-term research will also be necessary, beginning with the development of alternative conceptual models of the formation and maintenance of the landscape to guide a research program. Research on maintenance rather than evolution of the landscape should have higher priority because of its direct impact on restoration. Monitoring should be designed for the full range of flow conditions, including extreme events.

Overall, flows approximating historical discharges, velocities, timing, and distribution should be considered in restoration design, but quantitative flow-related performance measures are not appropriate until there is a better scientific understanding of the underlying science. At present, neither a minimum nor a maximum flow to preserve the landscape can be established.

Florida Bay Research Programs and Their Relation to the Comprehensive Everglades Restoration Plan (2002)

This report of the Committee on Restoration of the Greater Everglades Ecosystem (CROGEE) evaluated Florida Bay studies and restoration activities that potentially affect the success of the CERP. Florida Bay is a large, shallow marine system immediately south of the Everglades, bounded by the Florida Keys and the Gulf of Mexico. Some of the water draining from the Everglades flows directly into northeast Florida Bay. Other freshwater drainage reaches the bay indirectly from the northwest.

For several decades until the late 1980s, clear water and dense seagrass meadows characterized most of Florida Bay. However, beginning around 1987, the seagrass beds began dying in the western and central bay. It is often assumed that increased flows to restore freshwater Everglades habitats will also help restoration of Florida Bay. However, the CERP may actually result in higher salinities in central Florida Bay than exist presently, and thus exacerbate the ecological problems. Further, some percentage of the proposed increase in fresh surface-water flow discharging northwest of the bay will eventually reach the central bay, where its dissolved organic nitrogen may lead to algal blooms. Complicating the analysis of such issues is the lack of an operational bay circulation model.

The report notes the importance of additional research in the following areas: estimates of groundwater discharge to the bay; full characterization and quantification of surface runoff in major basins; transport and total loads of nitrogen and phosphorus from freshwater sources, especially in their organic forms; effects on nutrient fluxes of decreasing freshwater flows into the northeastern bay, and of increasing flows northwest of the bay; and the development of an operational Florida Bay circulation model to support a bay water quality model and facilitate analysis of CERP effects on the bay.

Science and the Greater Everglades Ecosystem Restoration: An Assessment of the Critical Ecosystems Study Initiative (2003)

The Everglades represents a unique ecological treasure, and a diverse group of organizations is currently working to reverse the effects of nearly a century of wetland drainage and impoundment. The path to restoration will not be easy, but sound scientific information will increase the reliability of the restoration, help enable solutions for unanticipated problems, and potentially reduce long-term costs. The investment in scientific research relevant to restoration, however, decreased substantially within some agencies, including one major Department of the Interior (DOI) science program, the Critical Ecosystem Studies Initiative (CESI). In response to concerns regarding declining levels of funding for scientific research and the adequacy of science-based support for restoration decision making, the U.S. Congress instructed the DOI to commission the National Academy of Sciences to review the scientific component of the CESI and provide recommendations for program management, strategic planning, and information dissemination.

Although improvements should be made, this report notes that the CESI has contributed useful science in support of the DOI's resource stewardship interests and restoration responsibilities in South Florida. It recommends that the fundamental objectives of the CESI research program remain intact, with continued commitment to ecosystem research. Several improvements in CESI management are suggested, including broadening the distribution of requests for proposals and improving review standards for proposals and research products. The report asserts that funding for CESI science has been inconsistent and as of 2002 was less than that needed to support the DOI's interests in and responsibilities for restoration. The development of a mechanism for comprehensive restoration-wide science coordination and synthesis is recommended to enable improved integration of scientific findings into restoration planning.

Regional Issues in Aquifer Storage and Recovery for Everglades Restoration: A Review of the ASR Regional Study Project Management Plan of the Comprehensive Everglades Restoration Plan (2002)

The report reviews a comprehensive research plan on Everglades restoration drafted by federal and Florida officials that assesses a central feature of the restoration: a proposal to drill more than 300 wells funneling up to 1.7 billion gallons of water a day into underground aquifers, where it would be stored and then pumped back to the surface to replenish the Everglades during dry periods. The report says that the research plan goes a long way to providing information needed to settle remaining technical questions and clearly responds to suggestions offered by scientists in Florida and in a previous report by the National Research Council.

Aquifer Storage and Recovery in the Comprehensive Everglades Restoration Plan: A Critique of the Pilot Projects and Related Plans for ASR in the Lake Okeechobee and Western Hillsboro Areas (2001)

ASR is a major component in the CERP, which was developed by USACE and the South Florida Water Management District (SFWMD). The plan would use the upper Floridian aquifer to store large quantities of surface water and shallow groundwater during wet periods for recovery during droughts.

ASR may limit evaporation losses and permit recovery of large volumes of water during multiyear droughts. However, the proposed scale is unprecedented and little subsurface information has been compiled. Key unknowns include impacts on existing aquifer uses, suitability of source waters for recharge, and environmental and/or human health impacts due to water quality changes during subsurface storage.

To address these issues, the USACE and SFWMD proposed aquifer storage recharge pilot projects in two key areas. The Committee on Restoration of the Greater Everglades Ecosystem charge was to examine a draft of their plans from a perspective of adaptive management. The report concludes that regional hydrogeologic assessment should include development of a regional-scale groundwater flow model, extensive well drilling and water quality sampling, and a multiobjective approach to ASR facility siting. It also recommends that water quality studies include laboratory and field bioassays and ecotoxicological studies, studies to characterize organic carbon of the source water and anticipate its effects on subsurface biogeochemical processes, and laboratory studies. Finally, it recommends that pilot projects be part of adaptive assessment.

Appendix B

Water Science and Technology Board and the Board on Environmental Studies and Toxicology

WATER SCIENCE AND TECHNOLOGY BOARD

CATHERINE L. KLING, *Chair*, Cornell University, Ithaca, NY
NEWSHA K. AJAMI, Stanford University, CA
JONATHAN D. ARTHUR, Florida Geological Survey, Tallahassee
FRANCINA DOMINGUEZ, University of Illinois, Urbana
DAVID A. DZOMBAK, Carnegie Mellon University, Pittsburgh, PA
WENDY D. GRAHAM, University of Florida, Gainesville
MARK W. LeCHAVALLIER, Dr. Water Consulting, LLC, Morrison, CO
MARGARET A. PALMER, SESYNC – University of Maryland, Annapolis, MD
DAVID L. SEDLAK, University of California, Berkeley
DAVID L. WEGNER, Jacobs Engineering, Tucson, AZ
P. KAY WHITLOCK, Christopher B. Burke Engineering, Ltd., Rosemont, IL

Staff

ELIZABETH EIDE, Director
LAURA J. EHLERS, Senior Program Officer
STEPHANIE E. JOHNSON, Senior Program Officer
M. JEANNE AQUILINO, Financial/Administrative Associate
BRENDAN R. McGOVERN, Research Assistant
CARLY BRODY, Senior Program Assistant

BOARD ON ENVIRONMENTAL STUDIES AND TOXICOLOGY

WILLIAM H. FARLAND, *Chair*, Colorado State University, Fort Collins
LESA AYLWARD, Summit Toxicology, LLP, Falls Church, VA
ANN M. BARTUSKA, Resources for the Future, Washington, DC
RICHARD A. BECKER, American Chemistry Council, Washington, DC
GERMAINE M. BUCK LOUIS, George Mason University
E. WILLIAM COLGLAZIER, AAAS, Washington, DC
DOMINIC M. DITORO, University of Delaware, Newark
DAVID C. DORMAN, North Carolina State University, Raleigh
GEORGE M. GRAY, The George Washington University, Washington, DC
R. JEFFREY LEWIS, ExxonMobil Biomedical Sciences, Inc., Annandale, NJ
R. CRAIG POSTLEWAITE, Department of Defense, Burke, VA
REZA J. RASOULPOURM, Corteva Agroscience, Indianapolis, IN
JOAN B. ROSE, Michigan State University, East Lansing
GINA M. SOLOMON, University of California, San Francisco
DEBORAH L. SWACKHAMER, University of Minnesota, St. Paul
JOSHUA TEWKSBURY, Future Earth, Boulder, CO
PETER S. THORNE, University of Iowa, Iowa City

Staff

CLIFFORD DUKE, Director
ELLEN K. MANTUS, Scholar and Director of Risk Assessment
RAYMOND A. WASSEL, Scholar and Director of Environmental Studies
SUSAN N.J. MARTEL, Senior Program Officer for Toxicology
LAURA LLANOS, Financial Associate
TAMARA DAWSON, Program Associate

Appendix C

Biographical Sketches of Committee Members and Staff

William G. Boggess, *Chair*, is professor and executive associate dean of the College of Agricultural Sciences at Oregon State University (OSU). Prior to joining OSU, Dr. Boggess spent 16 years on the faculty at the University of Florida in the Food and Resource Economics Department. His research interests include interactions between agriculture and the environment (e.g., water allocation, groundwater contamination, surface-water pollution, sustainable systems); economic dimensions and indicators of ecosystem health; and applications of real options to environmental and natural resources. Dr. Boggess previously served on the Oregon Governor's Council of Economic Advisors and the Board of Directors of the American Agricultural Economics Association, and he currently serves on the Board of the Oregon Environmental Council. He served on the State of Oregon Environment Report Science Panel and has been active in the design and assessment of the Oregon Conservation Reserve Enhancement Program. Dr. Boggess served as a member of the National Research Council (NRC) Committee on the Use of Treated Municipal Wastewater Effluents and Sludge in the Production of Crops for Human Consumption, and on the Committee on Independent Scientific Review of Everglades Restoration Progress (since 2008), serving as chair of the fourth committee. He received his Ph.D. from Iowa State University.

Mary Jane Angelo is a professor of law at the University of Florida's Levin College of Law and Director of the Environmental and Land Use Law Program. Her research areas focus on environmental law, water law, administrative law, biotechnology law, dispute resolution, pesticides law, law and science, and legal ethics. Prior to joining the faculty, Ms. Angelo served as an attorney in the U.S. Environmental Protection Agency's Office of General Counsel and as senior assistant general counsel for the St. Johns River Water Management District. She has served on several NRC committees, including the Committee on Ecological Risk Assessment under FIFRA and ESA and the Committee on Independent

Scientific Review of Everglades Restoration Progress (since 2010). She received her B.S. in biological sciences from Rutgers University and her M.S. and J.D. from the University of Florida.

Charley Driscoll (NAE) is university and distinguished professor in the Department of Civil and Environmental Engineering at Syracuse University where he also serves as the director of the Center for Environmental Systems Engineering. His teaching and research interests are in the area of environmental chemistry, biogeochemistry, and environmental quality modeling. A principal research focus has been the response of forest, aquatic, and coastal ecosystems to disturbance, including air pollution, land-use change, and elevated inputs of nutrients and mercury. Dr. Driscoll is currently a co-principal investigator of the National Science Foundation's Long Term Ecological Research Network's project at the Hubbard Brook Experimental Forest in New Hampshire. He is a member of the National Academy of Engineering and was a member of the NRC's Panel on Process of Lake Acidification and the Committees on Air Quality Management in the U.S. and the Collaborative Large-scale Engineering Analysis Network for Environmental Research (CLEANER). He has also served on the Committee on Independent Scientific Review of Everglades Restoration Progress since 2006. Dr. Driscoll received his B.S. in civil engineering from the University of Maine and his M.S. and Ph.D. in environmental engineering from Cornell University.

M. Siobhan Fennessy is the Jordan Professor of Biology and Environmental Science at Kenyon College, where she studies wetland ecosystems, particularly how wetland plant communities and biogeochemical cycles respond to human disturbances such as altered land use and factors associated with climate change. Her work has resulted in the development of biological assessment methods for wetlands that were recently employed in the National Wetland Condition Assessment effort led by the U.S. Environmental Protection Agency (EPA). She previously served on the faculty of the Geography Department of University College London and held a joint appointment at the Station Biologique du la Tour du Valat investigating human impacts to Mediterranean wetlands. She was a member of the EPA's Biological Assessment of Wetlands Workgroup, a national technical committee working to develop biological indicators of ecosystem condition. She recently co-authored a book on the ecology of wetland plants. Her current research focus is the alteration of ecosystem services that results from ecosystem degradation. Dr. Fennessy received her B.S. in botany and Ph.D. in environmental science from The Ohio State University. She served as a member of the National Academies' Committee to Review the St. Johns River Water Supply Impact Study.

Wendy D. Graham is the Carl S. Swisher Eminent Scholar in Water Resources in the Department of Agricultural and Biological Engineering and Director of the Water Institute at the University of Florida (UF), Gainesville. Her research focuses on integrated hydrologic modeling; groundwater resources evaluation and remediation; evaluation of impacts of agricultural production on surface- and groundwater quality; evaluation of impacts of climate variability and climate change on hydrologic systems; and stochastic modeling and data assimilation. In her role as director of the UF Water Institute she coordinates campus-wide interdisciplinary research, education, and outreach programs designed to develop and share new knowledge, and to develop and encourage the implementation of new technology and policy solutions needed to ensure a sustainable water future. She has a B.S. in environmental engineering from the University of Florida and a Ph.D. from the Massachusetts Institute of Technology.

Karl E. Havens is professor and director of Florida Sea Grant at the University of Florida. He has worked with Florida aquatic ecosystems and the use of objective science in their management for the past 23 years. His areas of expertise are in the fields of the response of aquatic ecosystems to natural and human-caused stressors, including hurricanes, drought, climate change, eutrophication, invasive species, and toxic materials, with particular attention to Florida's lakes and estuaries. Before joining the University of Florida, Dr. Havens was the chief environmental scientist at the South Florida Water Management District. He received his B.A. from SUNY Buffalo and his M.S. and Ph.D. from West Virginia University.

Fernando R. Miralles-Wilhelm is the executive director of the Cooperative Institute for Climate and Satellites, a cooperative institute between the University of Maryland and the National Oceanic and Atmospheric Administration, and a professor in the Department of Atmospheric and Oceanic Science at the University of Maryland. Dr. Miralles-Wilhelm specializes in hydrology and water resources engineering, with a particular focus on hydrology and climate interactions in the Everglades' vegetative ecosystems, which he has studyied for the past decade. Previously, he served on the faculty of Florida International University and the University of Miami. He received a mechanical engineering diploma from Universidad Simón Bolívar in Venezuela, an M.S. in engineering from the University of California-Irvine, and a Ph.D. in civil and environmental engineering from the Massachusetts Institute of Technology.

David H. Moreau is research professor, Department of City and Regional Planning, at the University of North Carolina at Chapel Hill. He recently completed a term as chair of the Curriculum for the Environment and Ecology. His research interests include analysis, planning, financing, and evaluation of water resource, water quality, and related environmental programs. Dr. Moreau is engaged in water resources planning at the local, state, and national levels. He has served on several NRC committees, including the Committee on New Orleans Regional Hurricane Protection Projects Review, the Committee on the Mississippi River and Hypoxia in the Gulf of Mexico, and the Committee on Independent Scientific Review of Everglades Restoration Progress (since 2006). Dr. Moreau recently completed 19 years as a member and 16 years as chairman of the North Carolina Environmental Management Commission, the state's regulatory commission for water quality, air quality, and water allocation. For his service to North Carolina he was awarded the Order of the Long Leaf Pine, the highest civilian award offered by the state. He received his B.S. and M.S. from Mississippi State University and North Carolina State University, respectively, and his Ph.D. from Harvard University.

Gordon H. Orians (NAS) is professor emeritus of biology at the University of Washington. Most of Dr. Orians's research has focused on behavioral ecology of birds and has dealt primarily with problems of habitat selection, mate selection and mating systems, selection of prey and foraging patches, and the relationship between ecology and social organization. Recently, his research has focused on environmental aesthetics and the evolutionary roots of strong emotional responses to components of the environment, such as landscapes, flowers, sunsets, and sounds. Dr. Orians has served on the Science Advisory Board of the U.S. Environmental Protection Agency and on boards of such environmental organizations as the World Wildlife Fund and the Nature Conservancy. He has also served on many National Academies committees, including the first Committee on Independent Scientific Review of Everglades Restoration Progress, the Committee on Cumulative Environmental Effects of Alaskan North Slope Oil and Gas Activities, and the Board on Environmental Studies and Toxicology. He is a member of the National Academy of Sciences and the American Academy of Arts and Sciences. Dr. Orians earned his B.S. in zoology from the University of Wisconsin and his Ph.D. in zoology from the University of California, Berkeley.

Denise J. Reed is a nationally and internationally recognized expert in coastal marsh sustainability and the role of human activities in modifying coastal systems with more than 30 years of experience studying coastal issues in the United States and abroad. Dr. Reed has served as a distinguished research professor in the University of New Orleans' Department of Earth and Environmental Sciences, and spent 5 years as chief scientist at The Water Institute of the Gulf. She has served on numerous boards and panels addressing the effects of human alterations on coastal environments and the role of science in guiding restoration, including the NRC Committee on Sustainable Water and Environmental Management in the California Bay-Delta, and she has been a member of the U.S. Army Corps of Engineer's Environmental Advisory Board and the National Oceanic and Atmospheric Administration (NOAA) Science Advisory Board. Dr. Reed received her B.S. in geography from Sidney Sussex College and her M.A. and Ph.D. from University of Cambridge.

James E. Saiers is professor of hydrology, the associate dean of academic affairs, and professor of chemical engineering at the Yale School of Forestry and Environmental Studies. Dr. Saiers studies the circulation of water and the movement of waterborne chemicals in surface and subsurface environments. One element of his research centers on quantifying the effects that interactions between hydrologic and geochemical processes have on the migration of contaminants in groundwater. Another focus is on the dynamics of surface water and groundwater flow in wetlands and the response of fluid flow characteristics to changes in climate and water management practices. His work couples field observations and laboratory-scale experimentation with mathematical modeling. Dr. Saiers was a member of the NRC's Committee on Independent Scientific Review of Everglades Restoration Progress and chaired the Committee to Review the Florida Aquifer Storage and Recovery Regional Study Technical Data Report. Additionally, he served as a member of the Hydraulic Fracturing Research Advisory Panel of the Environmental Protection Agency Science Advisory Board. He earned his B.S. in geology from the Indiana University of Pennsylvania and his M.S. and Ph.D. in environmental sciences from the University of Virginia.

Eric P. Smith is a professor in the Department of Statistics at the Virginia Polytechnic Institute and State University. Dr. Smith's research focuses on multivariate analysis, multivariate graphics, biological sampling and modeling, ecotoxicology, data analytics, and visualization. He teaches courses in biological statistics, biometry, consulting, data mining, and multivariate methods. His courses focus on extracting information from large data sets and on analyzing and solving problems through fast algorithms, accurate models, evolving statistical methodology, and quantification of uncertainty. He is the former director of the Computational Modeling and Data Analytics Program. He earned his B.S. from the University of Georgia and his M.S. and Ph.D. from the University of Washington.

Denice H. Wardrop is a senior scientist and professor of geography and ecology at The Pennsylvania State University. She also serves as the director of its Sustainability Institute and as assistant director of Penn State Institutes of Energy and the Environment. Her research focuses on theoretical ecology, anthropogenic disturbance and impacts on aquatic ecosystem function, ecological indicators, and ecosystem condition monitoring and assessment. Dr. Wardrop is the Pennsylvania Governor's Appointee to the Chesapeake Bay Program's Science and Technical Advisory Committee and previously served as its chair. She also directs the Mid-Atlantic Wetlands Workgroup. She has a B.S. in systems engineering from the University of Virginia, an M.S. in environmental sciences from the University of Virginia, and a Ph.D. in ecology from the Pennsylvania State University.

Greg D. Woodside is the executive director of planning and natural resources at Orange County Water District. Mr. Woodside has 25 years of experience in water resources management and hydrogeology. He is a registered geologist and certified hydrogeologist in California, and oversees the Planning and Watershed Management Department and the Natural Resources Department at the Orange County Water District. Staff in these departments prepare the District's environmental documents, permit applications, groundwater management plan, and long-term facilities plan, and conduct the District's natural resource management, watershed planning, and recharge planning. In particular, he has evaluated conjunctive use and aquifer storage and recovery projects in the Orange County Groundwater Basin, Central Basin, and San Gabriel Basins, including projects that would recharge up to 50,000 acre-feet per year of recycled and imported water. Mr. Woodside previously served on the National Academies' Committee to Review the Edwards Aquifer Habitat Conservation Plan. He holds a B.S. in geological sciences from California State University, Fullerton, and an M.S. in hydrology from the New Mexico Institute of Mining and Technology.

STAFF

Stephanie E. Johnson, study director, is a senior program officer with the Water Science and Technology Board. Since joining the National Research Council in 2002, she has worked on a wide range of water-related studies, on topics such as desalination, wastewater reuse, contaminant source remediation, coal and uranium mining, coastal risk reduction, and ecosystem restoration. She has served as study director for 20 committees, including the Panel to Review the Critical Ecosystem Studies Initiative and all seven Committees on Independent Scientific Review of Everglades Restoration Progress. Dr. Johnson received her B.A. from Vanderbilt University in chemistry and geology and her M.S. and Ph.D. in environmental sciences from the University of Virginia.

David J. Policansky is a scholar and director of the Program in Applied Ecology and Natural Resources of the Board on Environmental Studies and Toxicology. He earned a Ph.D. in biology from the University of Oregon. Dr. Policansky has directed approximately 35 National Research Council studies, and his areas of expertise include genetics; evolution; ecology, including fishery biology; natural resource management; and the use of science in policy making.

Brendan R. McGovern is a research assistant with the Water Science and Technology Board. Mr. McGovern has contributed to a number of studies and activities, on topics such as municipal water supply, aquifer storage and recovery, community-based flood insurance, ecosystem restoration, and coastal risk reduction. He previously worked and interned with the American Association for the Advancement of Science and the Stimson Center on international water security issues. He earned his B.A. degrees in political science and history from the University of California, Davis.